AF553445

ADVANCED AGRO-TECHNOLOGY

ADVANCED AGRO-TECHNOLOGY

Ashok Kumar
Dept. of Zoology
Bundelkhand University
Campus Department
Jhansi

DISCOVERY PUBLISHING HOUSE PVT. LTD.
NEW DELHI-110 002

Published by:
Namit Wasan
DISCOVERY PUBLISHING HOUSE PVT. LTD.
4383/4B, Ansari Road, Darya Ganj
New Delhi-110 002 (India)
Phone : +91-11-23279245; 23253475; 43596065
E-mail : discoverybooksindia@gmail.com
discoverypublishinghouse@gmail.com
namitwasan9@gmail.com
web : www.discoverypublishinggroup.com

Edition: **2020**

ISBN: 978-81-8356-484-7

Advanced Agro-Technology

Printed at:
Infinity Imaging Systems
Delhi

Preface

Our agricultural scene is a changing rapidly since the mid-sixties when we experienced the 'Green Revolution'. It was followed by the 'White Revolution', which brought in the flood of milk in the country. Now, we are on way to 'Yellow Revolution' i.e. revolution in fruit production. An important fact behind these revolutions is the economic consideration. Farmers have been adopting the new technology in agriculture and related fields mainly because of these innovations pay more all compared to their tradition crops.

Agriculture has become more market oriented. The considerations of cost of production, income and profitability are summing significance. Thus, the Agri-economics has come in. The farm management is the field closely related to agri-ecomnomics. There is no doubt that farmers adopting new technology are quite conscious about the management of their farm. However, there is general apathy about the farm management amongst the farmers, Agri-scientists and the policy makers. Being a relatively young since, no much attentions is post to it. There may be some sound reasons for this. However, with changing scenario in agriculture, the need for guiding farmers in management of their farms assumes significance. In any business proposition, management has to play a significant role. In other sectors of the economy, 'management' has become a catchword. In recent years, the irritations giving training in 'management' are increasing by leaps and bounds.

There is good demand for 'managers' trained in these institutions. Now, the technical persons having engineering and other vocational background are turning towards MBA and allied courses. However, with rare exceptions, there are no such institutions for training persons in farm management. These are institutions for guiding farmers in this subject. One of the reasons for this may be that our farming is relatively a small-scale and a scattered one. However, looking to the changes that are taking place, in agriculture sector, there is need for guiding the farmers in this economic aspect.

The subject matter covers several aspects such as decision-making, farm planning, resource planning (land, labour, capital), farm costing, farm accounting, marketing, analysis of case studies of successful farmers.

The use of efficiency measures to evaluate the enterprise is also an important aspect of farm management. In the recent years, the statistical tools such a production function, linear programming are being used to get an idea about the resource. We in an enterprise and for getting an optimum plan for a given farm taking into account the resources and the limitations in which it is being operated. In the near future the farm management is likely to assume more and more significance.

—Author

Contents

1

Introduction

GREEN REVOLUTION

Since the Green Revolution, India has been growing steadily in terms of agricultural productivity and growth. Modern farming practices and inclusive technologies have been implemented in many parts of rural India to foster rural growth. Wireless communication networks and GIS-based agro-software technology are reaching rural India giving them access to vital and updated information on weather, farming technologies, fertilisers, livestocks, commodity prices and stock markets.

Despite the usage of advanced ICT tools in many parts of rural India, there are several villages where advanced farming technologies, and interactive communications networks haven't reached yet. Access to these technologies will help secure livelihood of many farmers and consequently build a stronger nation and economy.

Building up on the achievements of the first eAgriculture conference, eAgriculture India 2008 proposes to take forward the momentum created last year. eAgriculture India 2008 will reinitiate a knowledge revolution in Indian agriculture and agronomy by deliberating on the unsolved issues and by exploring ways through which current practices can be made more efficient.

The Agriculture track aims to create new linkages and strengthen existing bonds among the various stakeholders and enhance their capacity to find working solutions to the issues being faced by farming communities around the world. The conference also aims to identify the areas that need more attention and to devise more suitable methods of including the information have-nots into the knowledge revolution that this initiative would bring about.

New initiatives being taken up by both Public and Private stakeholders – often entering into a joint venture – have the potential to revolutionise the information delivery mechanisms and make them more democratic, accessible and inclusive in nature. The fact that communication technology is becoming increasingly economical has brought about a sea change in the way information can be disseminated.

There are many other ways through which ICT can influence the lives of countless farmers around the world and usher in a revolution in global agronomy especially with regards to agriculture in the developing nations. The event seeks to identify many more ways through which the farming communities can benefit from the technologies that are at our disposal. Participants are encouraged to present their papers during the event for which we are accepting abstracts right now.

FARM MANAGEMENT

Our agricultural scene is changing rapidly since the mid-sixties when we experienced the 'Green Revolution'. It was followed by the 'White Revolution', which brought in the flood of milk in the country. Now, we are on way to 'Yellow Revolution' i.e. revolution in fruit production. An important fact behind these revolutions is the economic consideration. Farmers have been adopting the new technology in agriculture and related fields mainly because of these innovations pay more all compared to their tradition crops. Agriculture has become more market oriented. The considerations of cost of production, income and profitability are summing significance. Thus, the Agri-economics has come in. The farm management is the field closely related to agri-ecomnomics. There is no doubt that farmers adopting new technologies are quite conscious about the management of their farm. However, there is general apathy about the farm management amongst the farmers, Agri-scientists and the policy makers. Being a relatively young since, no much attention is post to it. There may be some sound reasons for this. However, with changing scenario in agriculture, the need for guiding farmers in management of their farms assumes significance. In any business proposition, management has to play a significant role. In other sectors of the economy, 'management' has become a catchword. In recent years, the irritations giving training in 'management' are increasing by leaps and bounds. There is good demand for 'managers' trained in these institutions. Now, the technical persons having engineering and other vocational background are turning towards MBA and allied courses. However, with rare exceptions, there are no such institutions for training persons in farm management. These are institutions for guiding farmers in this subject. One of the reasons for this, may be that our farming is relatively a small-scale and scattered one. However, looking to the changes that are taking place, in agriculture sector, there is need for guiding the farmers in this economic aspect. The subject matter covers several aspects such as decision-making, farm planning, resource planning (land, labour, capital), farm costing, farm accounting, marketing, analysis of case studies of successful farmers. The use of efficiency measures to evaluate the enterprise is also an important aspect of farm management. In the recent years, the statistical tools such a production function, linear programming are being used to get an idea about the resource. We in an enterprise and for getting an optimum plan for a given farm taking into account the resources and the limitations in which it is being operated. In the near future the farm management is likely to assume more and more significance.

Decision-making in any business largely depends on the nature of business. To understand decision-making in farming, it is necessary to know the nature of Indian farming.

NATURE OF INDIAN FARMING

Indian farming is basically a subsistence farming. But due to new technological developments in all the areas of production, productivity of crops and live-stock has increased substantially, total production has increased several times and marketable surpluses with the farmers have also increased. As a result farming has assumed commercial proposition.

Indian farming is a diversified and mixed type of farming. The farmers grow 5-6 crops or even more on their farms.

To meet their family requirements.

To suit their soils on the farms and considering availability of resources including irrigation and market facilities. They also follow mixed farming, that is they keep few milch animals and follow dairy and also undertake sheep or goat keeping poultry keeping, etc. to supplement incomes from crop production. Therefore decision-making on such farms is more difficult and complex than on specialised farming.

Indian farming is a family type farming. Farm family provides major part of labour and also provides some capital.

Role of Farmers:

Indian farmer plays following four rules simultaneously:

He is an entrepreneur

He is a Manager

He is a Financer

He is a Labourer or Worker

He takes decisions in two capacities as an entrepreneur and a manager.

Area of decision-making:

There are four major areas of decision-making:

Production of crops and live-stock.

Acquiring inputs.

Obtaining credit.

Marketing his produce.

Production: The decisions related to production activity are:

What to produce? i.e. selection of crops and live-stock activity. Whether to have dairy, goat keeping or poultry or their combination.

How to produce? Adoption of suitable technology, whether to continue old technology or use new technology or combination of both.

How much to produce? To consider family requirements in respect of foodgrain, vegetable, spices, fruits, milk, etc and also think of producing marketable surpluses. In that case, think of demand supply position, availability of market and prevailing prices. This will also apply to production of cash/commercial crops.

Acquiring inputs: In modern commercial farming, wide range of inputs are required. They include seeds of high yielding and hybrid varieties, fertilisers, pesticides, plant growth regulators (hormones), improved implements and machinery, etc. There are commercial and specialised agencies dealing in these inputs. The farmer has to take decision and make choices relating to:

From whom to buy?

At what price to buy?

How much to buy?

Whether to buy on cash or on credit?

Obtaining credit: Although the farm family supplies some finance, it is not adequate considering larger financial requirements of modern agriculture. Therefore, farmer has to borrow from outside sources. In this connection he has to take following decisions. There is risk in borrowing.

How much to borrow?

From whom to borrow? There are some alternative sources viz. Moneylenders, banks, government incentive schemes, friends, relatives, traders, etc. He has to make right choice:

At what rate to borrow?

What security/mortgage to offer?

How repayments to be made?

How to face risk situation?

Marketing the produce: When surplus produce is obtained or created, its disposal or marketing advantageously becomes necessary. Following decisions are involved in the marketing of farm produce.

Where to sell? This is the most important decision because in rural areas big, wholesale and organised markets are limited. As far a individual farmer is concern, his produce is of small quantity, he has no transport facility, there are no good roads to go to distant markets. The farmer has to make choice within limited alternatives.

Whom to sell? There are various types of agencies such as villager trader, wholesale cum commission agents, co-operatives, government (for selected commodities). He has to make choice from among them.

When to sell? Immediately after harvest when prices are the lowest or wait for better prices. He is also constrained by cash requirements for family expenditure. He has to take decision under most unfavourable situation.

At what price to sell ? Here the position of farmer is very vulnerable. He has almost no choice. Prices in organised markets (Regulated markets) are fixed by open auction and he has to accept the price quoted by the highest bidder even this price may not be remunerative. In unorganised markets, prices are fixed by traders and farmer is hardly consulted. He has very limited scope for decision-making.

DECISION-MAKING

Decision-making is an art as well as science. It is more a brain work. The quality of decision-making is dependent on level of formal education, knowledge of the subject and experience. It is a continuos process.

The responsibility of decisions is borne by the decision maker in terms of rewards and punishments for taking right or wrong decisions. If he takes right decisions, production increases, income and profit increases, and his family welfare improves. This is a reward for him. If he takes wrong decisions the level of production goes down, there is no profit or low profit and low family welfare. He gets self-punishment.

2

Integrated Natural Resource Management Research

INRM

Support for research on natural resource management appears to be approaching a crisis. Increasingly, questions are being raised as to whether this research can deliver the goods. Some feel that it is more concerned with definitions and purity of process than with results. Research on natural resource management must demonstrate its ability to benefit large numbers of poor people across large areas within sensible time frames. The easy assumption that such work is inherently site specific must be overturned. Put simply, we must meet the challenge of accelerating the use of natural resource management practices that improve human well-being.

Integrated natural resource management (INRM) research can meet this challenge. Decentralised initiatives, supported by effective institutions and guided by suitable information management tools, can lead to the widespread use of suitable management options from INRM research. This, in turn, can improve agroecosystem productivity and resilience, thereby helping achieve the goals of poverty alleviation, food security, and environmental protection. Behind this is the realisation that policies, people's behaviour, natural resource management practices, biophysical processes, and system outcomes are linked in cause-and-effect relationships (Fig. 2.1). Specifically:

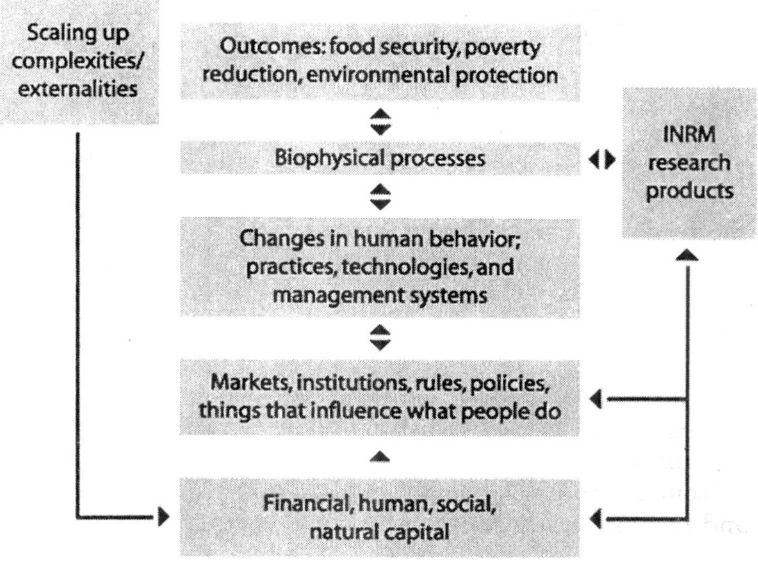

Fig. 2.1: Integrated natural resource management research furthers the goals of the Consultative Group on International Agricultural Research (CGIAR): food security, poverty reduction and environmental protection.

- policies, organisations, institutions, and rules affect the behaviour of communities and individual farm families;
- people's behaviour includes the selection and adoption of natural resource management practices;
- these practices affect plant and animal growth and biophysical processes; and
- biophysical processes result in outcomes that have consequences for incomes, food security, and resource conservation.

This chapter discusses some of the concepts involved in and procedures for generalising and propagating the results of natural resource management research ("scaling out"), with a few forays into the area of externalities and scale of analysis ("scaling up"). It features examples of several methods and tools for accelerating the scale of geographical coverage and impact of INRM practices. Most examples are drawn from collaboration between the Centro Internacional de Mejoramiento de Maiz y Trigo (CIMMYT), known in English as the International Maize and Wheat Improvement Center, and research partners in South Asia, southern Africa, and Mesoamerica. Methods and tools illustrated include site similarity analysis through geographic information systems (GIS), the linking of simulation models with GIS, and farmer and land type categories. The selection of examples is illustrative and does not aim to be comprehensive.

These examples show the tools being used in the context of a problem-solving process that harnesses cause-effect links among policies and institutions, farm-level practices, plant and animal growth, biophysical processes, and impacts and outcomes. Strengths and weaknesses of the different methods and tools are discussed.

Finally, it is argued that these tools are most useful when they provide information in the context of a bottom-up learning process to a wide range of stakeholders who need this information to make decisions. They should never be used for the mere mechanical extrapolation or replication of particular practices.

A PROBLEM-SOLVING APPROACH

Research on integrated natural resource management (INRM) must be capable of solving problems (or seizing opportunities) in ways that improve livelihoods for the poor while conserving resource quality and protecting the environment. Understandably, INRM researchers may wish to apply a problem-solving approach (Tripp 1991). Within a problem-solving process, we can distinguish among problem sets, causes, intervention points, and measurement tools.

Problem sets are situations in which agroecosystem performance, i.e., the processes that affect the resource quality or the environment, is unsatisfactory. Examples include low agroecosystem productivity, excessive resource degradation and environmental pollution, low levels of environmental services, low agroecosystem biodiversity, reductions in soil fertility, unsatisfactory water quality for consumers, and excessive greenhouse gas emissions. These problems can be characterised in terms of their costs and consequences, spatial and temporal incidence, and pace of change. They can be recognised and defined by farmers, communities, non-governmental organisations (NGOs), scientists, and/or policy makers.

Causes are the factors that drive or contribute to problem sets. Typically, many causes at several levels are at work. Causal chains can be long and complex, linking policies, institutions, farmer or community behaviour, biophysical processes, and their consequences for livelihoods and the environment. In other words, policies and institutional arrangements affect people's behaviour, people's behaviour affects plant and animal growth and biophysical processes, and biophysical processes result in outcomes that cause changes in system productivity and resource and environmental quality.

Chains of cause and effect typically link different scales of analysis. For example, regional policies on the burning of crop residues may influence mulch management at the farm level, affecting soil water and organic matter levels and fractions and rates of erosion at the plot level, with consequences for water quality in the watershed as well as for crop yields and family incomes at the farm level.

Intervention points are opportunities for addressing the problem set. They are not restricted to new farm-level technologies; they may also include changes in policies and institutional arrangements, e.g., rules governing community forest management. However, policy change as an intervention is most effective when cause-and-effect relationships are clear, that is, when there is a reasonable likelihood that a change in policies or institutions will modify farmer or community behaviour in ways that lead to desired changes in biophysical processes, system productivity, and environmental and resource quality. Interventions, then, can be at any level of analysis: plot, farm, community, watershed, or region. They may be developed by farmers via farmer experimentation, by scientists, by policy makers, or by the private sector. Early successful interventions have been referred to as "sparks" (Consultative Group on International Agricultural Research 2000).

For example, a problem set may revolve around the siltation of the lowland irrigation infrastructure, leading to substantial productivity losses and heavy public investment in renovation. Causes may include heavy erosion from upland areas driven by policies that encourage communal livestock grazing of crop residues, thus reducing incentives to use these residues as a soil cover. An intervention point might feature policy changes to foster modifications in grazing practices that encourage the use of crop residues as a soil cover mulch to reduce erosion and ameliorate the original problem of siltation.

Finally, measurement tools allow us to understand cause-and-effect links, trace and even anticipate the consequences of interventions, and understand biophysical processes at any scale of analysis. Indicators of sustainability fall into this area, as do most modeling approaches. In this vein, ecosystems analysis provides an analytical framework that makes it easier to understand the consequences of changes in both short- and long-term states at a range of scales. The processes can be linked conceptually within a framework (see Fig. 2.2), and the effects of given scenarios can be quantified using simulation models linked to spatial and temporal databases through GIS.

Of course, most models still need to be refined in the critical areas of edaphic and pest (insects, pathogens, and weeds) interactions and constraints. Ecosystems analysis can provide two critical services at relatively minor cost: (1) assessment of both genetic and environmental productivity and sustainability and (2) a framework for impact assessment and the definition of problem-cause relationships, especially those involving biophysical processes, and how those relationships affect system productivity and sustainability. INRM will fail if we do not have a problem focus and include plenty of work to identify intervention points; we cannot simply conduct academic work on measurement tools.

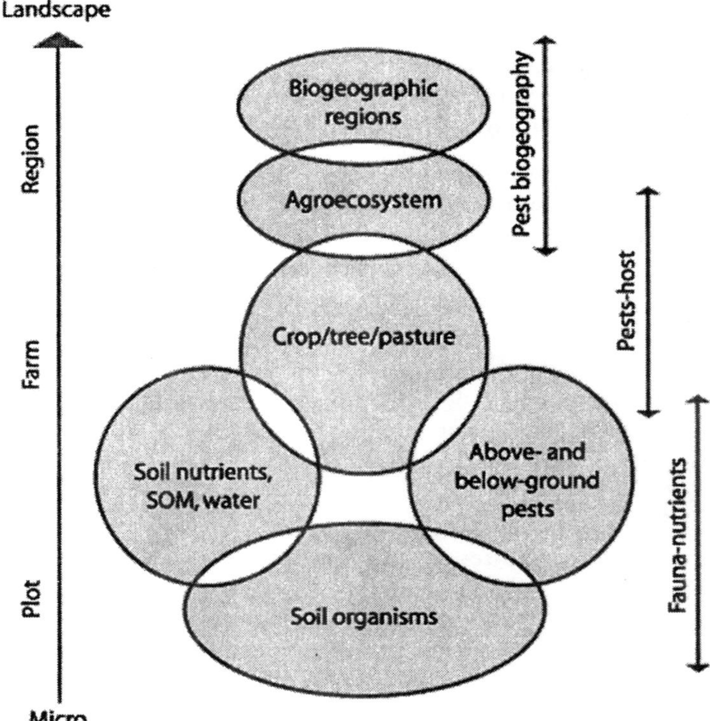

Fig. 2.2: Biophysical processes at different scales of analysis. SOM stands for "soil organic matter".

SIMPLE INTERVENTIONS IN COMPLEX SYSTEMS

Natural resource management practices as implemented by resource managers such as farmers, communities, fishers, and forest dwellers are typically complex. Rules governing the use of land and water resources or forest or fishery stocks are usually complicated and difficult for outsiders to understand. However, intervention points, including new technologies or practices for resource use, can be relatively simple. Interventions are usefully seen as options or alternatives for exploration by resource users, who can best judge the attractiveness of an option by testing it under local circumstances.

However, even for simple interventions the consequences of widespread adoption can be hugely complicated. The introduction of relatively simple options can significantly change farming or resource management systems and their accompanying biophysical processes and system outcomes.

For example, farmers who deal with irrigated crop systems use complex practices to manage soil fertility and water quantity and quality. These include managing crop residues, fertilisers, and farmyard manure; arranging for biomass transfer from outside the farm; choosing alternative fuels for household use; deciding among alternative uses for canal and tubewell water; making decisions related to the timing and frequency of irrigation; and selecting crops for well-drained vs. poorly drained areas, among other things (Fujisaka et al. 1994). However, the introduction of a relatively simple practice such as zero-tillage crop establishment can improve the timeliness of sowing, increase the efficiency of water and nutrient use, reduce water pumping, stop groundwater depletion, reduce fuel use, drastically lower carbon emissions, change crop rotations to take advantage of the earlier grain-crop sowing, and change soil chemistry and soil health via new rotations (Hobbs and Morris 1996). Some of these consequences, e.g., changes in the quality and quantity of groundwater, may become apparent only at higher scales of analysis.

A good understanding of ecological, biophysical, economic, and social processes is needed to anticipate, model, assess, and manage such changes. Otherwise, farmers and scientists alike can only react to changes as they unfold.

THE NOTION OF SCALE

The role of integrated natural resource management (INRM) in "delivering the goods," that is, in fostering improvements in the livelihoods of large numbers of the poor, is often referred to as scaling out. This phrase conceals as much as it clarifies, because the notion of "scale" is perceived in many different ways, among them:

- *scale of analysis:* from plant to plot to farm to watershed to region;
- *scale of intervention point:* high-level interventions such as policy changes, adjustments in institutional arrangements or property rights, and the fostering of collective action vs. lower-level interventions such as farmer experimentation or extension for specific practices;
- *scale of investment in intervention strategies:* small vs. large investments in extension, farmer experimentation programmes, or efforts to provide information to policy makers;
- *scale of community empowerment:* the number of communities able to undertake their own research and adaptation through processes for local learning;
- *scale of geographical coverage of an INRM practice:* whether it is limited to a village or watershed or has attained regional or national relevance;
- *scale of impact:* for example, the extent to which desirable outcomes, e.g., improved system productivity and resource quality, have been achieved through INRM research.

In principle, these scales are linked. Greater impacts are generated from higher levels of investment in suitable intervention strategies, or from more efficient use of these investments through greater reliance on community empowerment, leading to expanded geographical coverage of suitable practices.

This paper focuses on ways to augment and accelerate the scale of geographical coverage and impact of INRM research. It emphasizes efficiency and effectiveness in generalising and propagating research results through the replication, dissemination, and adaptation of technologies or practices. These technologies may comprise plausible promises, malleable prototypes, or well-defined practices. If INRM research products are not scaled out, we will have failed in our goal of contributing to poverty alleviation, food security, and environmental protection.

Sometimes, though, to augment and accelerate INRM research impacts, we must also assess and manage positive or negative externalities, unexpected complexities, or unintended consequences that emerge at higher scales of analysis from the widespread adoption of new resource management practices. This is because consequences that emerge only at higher scales of analysis may either reinforce or undermine the desired outcomes. For example:

- improved efficiency of water use at the plot level may not, in fact, lead to improved water use at the level of the whole irrigation system;
- changes in land use or crop management on hillsides may improve or possibly downgrade the quantity and quality of water available to downstream users;
- more efficient fishing practices used by one person may destroy fish stocks if used by everyone;
- local rules and incentives may be undermined by regional or national policies; and
- institutions that seek to control rather than manage biophysical processes may not foster adaptive capacity, possibly exacerbating rather than solving problems.

Effective scaling out, then, also requires attention to the other notions of scale: scale of analysis, scale of intervention point, and scale of community empowerment.

SCALING OUT AND COMMUNITY EMPOWERMENT

Much current thinking on scaling out integrated natural resource management (INRM) research steers clear of the notion of spatial extrapolation of specific practices. Rather, the emphasis is on community empowerment and scaling out as a learning process. A recent workshop report (Consultative Group on International Agricultural Research 2000) describes this well.

It is not technologies that are scaled up, but processes and principles behind the technologies/innovations. This is consistent with the belief that scaling out is not just replication but adaptation and learning that is flexible and interactive... Scaling out is really about people—of communicating options to people, of a balance between introducing options and involving farmers' ability to adapt to changing contexts...Scaling out as a development process rejects the cookie cutter approach. [It]... achieves large numbers and wide area coverage through multiplication with adaptation...

We agree with these conclusions. Bottom-up farmer experimentation and community empowerment are fundamental to scaling out INRM practices. However, these bottom-up approaches will be more effective when their outcomes are widely shared. Surely farmer experimenters are likely to be interested in trying out exciting practices developed in similar communities facing similar problems.

Although scaling out is largely a bottom-up process whereby research outcomes are widely shared, our experience suggests that the use of information technologies such as the methods and tools described below can help "smarten" and focus the process:

ELEMENTS OF SCALING OUT

What can be done to foster the effective scaling out of suitable natural resource management practices? We suggest some activities below, several of which involve improvements in human and social capital.

- *Generate more attractive products.* Regardless of how it is done, scaling out is easier when practices are less risky and more profitable, and meet other resource management objectives. Participatory research increases the chance of identifying attractive options;
- *Balance supply-driven approaches with resource-user demands.* Demands from resource users must influence the kinds of resource management options developed through research and the kinds of options to be scaled out. However, they cannot express a demand for practices with which they are unfamiliar. Scaling out, then, must

include ways for users to become familiar enough with new options to judge their attractiveness under local conditions;

- *Use feedback to redefine the research agenda.* As information accumulates on technology performance and attractiveness and how policies and institutions influence them, integrated natural resource management research can and should be adjusted accordingly;
- *Encourage support groups and networks for information sharing.* Community groups, cross-community networks, alliances of networks, study tours, and scientific exchanges can all help resource users as well as scientists better understand the performance of alternatives and options under different conditions;
- *Facilitate negotiation among stakeholders.* With multiple-function, multiple-user resources, trade-offs in resource use may lead to conflicts among stakeholders. Negotiation and conflict management among stakeholders may be helpful in resolving conflicts and encouraging the use of suitable practices;
- *Provide information of use to those who are establishing policies and developing institutions.* Scientists can provide helpful information for policy formulation and institutional development. For example, if adaptable institutions are needed to review new resource management practices, this should be made clear. Policy makers may welcome new information on how resource management practices can help them meet economic and social goals. New policies and institutions can influence human behaviour, including technology adoption;
- *Make sensible use of information management tools such as GIS and modelling.* When practices that raise agroecosystem productivity, improve resource quality, and ameliorate environmental consequences are discovered or developed, there is an understandable interest in seeing that these practices or their adaptations are used more

widely. Adding a spatial dimension to the problem-solving process can help make this happen. This results from the simple recognition that practices may be equally attractive to different farmers or farming communities that face similar problem sets, are driven by similar causes, and are governed by similar factors with regard to adoption behaviour. This is not a plea for top-down mechanical extrapolation of technology; rather, it is the recognition that stakeholders can use the information provided by spatial analysis when making decisions.

For example, in a certain community, a green manure cover crop may smother weeds, free up labour, improve water use efficiency, reduce the need for external inputs, raise yields, and improve farm family livelihoods. Research may suggest that this practice is most attractive in locations where the cover crop is climatically adapted, soil fertility is within a certain range, land use intensity is low (allowing a cover crop/grain crop rotation), and marketing margins are high (making external input use unprofitable). Spatial analysis that combines data on the climate, soils, population density, crop distribution, and transport infrastructure can identify large areas in other communities that might benefit from this practice. This outcome can be shared with NGOs, research and extension institutions, farmer groups, and policy makers for use as they see fit. This may encourage NGOs or farmer groups to experiment with and adapt the practice, or at least to evaluate its attractiveness under local conditions.

INFORMATION MANAGEMENT TOOLS FOR SCALING OUT: EXAMPLES

The following sections provide examples of information management tools of potential use in scaling out integrated natural resource management (INRM) practices. Most examples are drawn from CIMMYT's collaboration with a range of partners in South Asia, southern Africa, and Mesoamerica. The selection of examples is illustrative, not comprehensive. The methods and tools discussed include site similarity analysis through GIS, the linking of simulation models with GIS, and the use of farmer and land type categories. Although, in most instances, the tools and methods show considerable promise for use in scaling out INRM practices, on-the-ground experience remains insufficient. The strengths and weaknesses of these methods and tools are presented in Table 2.1.

Table 2.1: Relative strengths and weaknesses of tools and methods for scaling out

Tool or method	Strength	Strength
Site similarity analysis	Simple tools available Conceptually accessible	May over simplify Criteria for similarity often subjective
Interfacing GIS with models	Allows examination of time trends, including climatic risk can express outputs in terms of specific variables of interest of stakeholders	Dependent on equality of model requires specialists to implement
Land type and farmer categories	Outputs conceptually accessible outputs suitable for use by extension workers and farmer experimenters	Outputs possibly too subjective labour intensive data acquisition may ignore interactions across land types within a household
Participatory extension e.g., whole family training	Outputs readily accessible to farm families Can be scaled up in terms of organisational capacity required for implementation	Deals only with the family as a unit, does not extend to collective action at the community level does not have an explicit spatial dimension

SITE SIMILARITY ANALYSIS THROUGH GIS

A recurring question in efforts to scale out promising interventions is how a practice developed at one location will perform over a broader range of environments. Geographic information systems (GIS) can address such concerns, allowing scientists to share relevant results with colleagues elsewhere, to find new sites for testing and adapting discoveries, and to design more effective research programmes. One simple

GIS-based approach is to identify areas that are similar to a given location, using criteria relevant to the problem at hand (Corbett et al. 1999).

To identify regions suitable for the introduction and adaptation of wheat production practices that might show promise for conditions in Bolivia, a GIS was used to identify sites similar to key research locations in the country's two major wheat system environments (Hodson et al. 1998). In the highlands, wheat is grown on summer rains in numerous valleys and small plateaus. In the eastern lowlands, the crop is sown on residual soil moisture as temperatures drop and become more favorable for wheat. Zones of similarity were defined using the GIS-based Spatial Characterisation Tool (Corbett and O'Brien 1997) by specifying the latitude and longitude of a given research site and then selecting criteria for similarity based on ranges of precipitation, potential evapotranspiration, and temperature. For the highlands, zones were based on the favorable 5-month growing period, and for the lowlands, the coolest quarter was used.

There were scattered zones of similarity in the highlands of Bolivia, Peru, Colombia, and Venezuela. Extending the analysis to Mexico, Central America, and Africa resulted in the identification of additional areas with similar climates, notably in Mexico and Ethiopia (*See fig. 2.3 on page 22*). For lowland sites, the largest regions outside of Bolivia were in two substantial but disjunct areas of eastern and southwestern Brazil. To extend the analysis to a complete farming system scenario, similarity zones for the rainy season were identified to account for the times when crops such as maize, cotton, and soybean were normally sown and harvested. This allowed researchers to narrow regions of similarity to a single area in eastern Brazil (*See fig. 2.4 on page 23*).

This specific analysis has yet to be tapped to scale out INRM practices. However, it has now become clear that researchers and farmer experimenters in the defined areas of Bolivia, Brazil, Mexico, and Ethiopia are addressing similar problem sets with similar interventions, and would benefit from sharing research information and results.

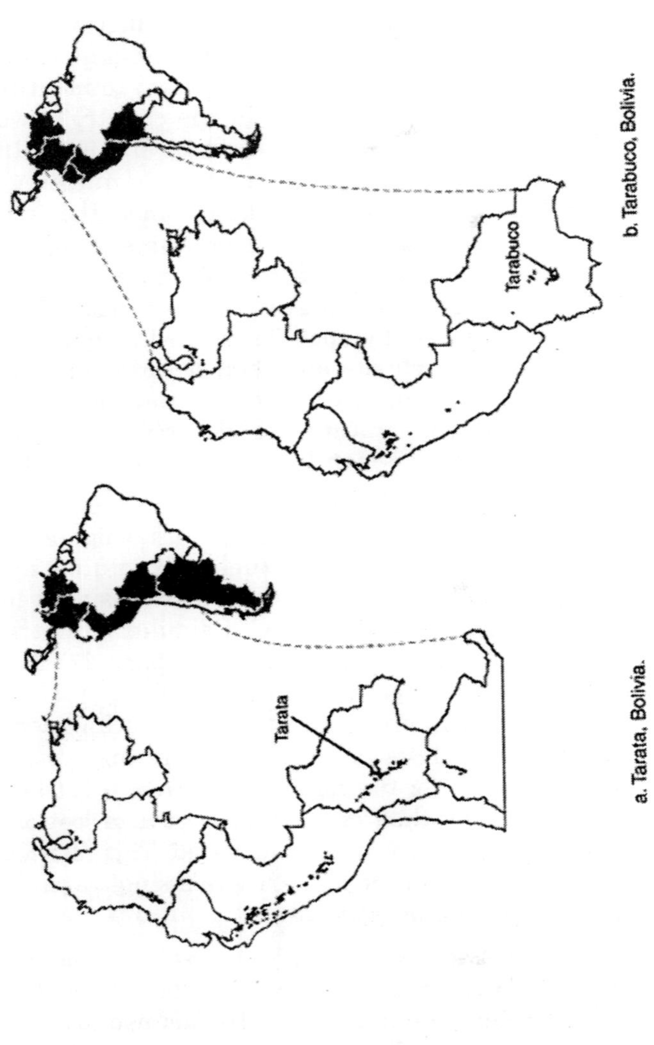

Fig. 2.3: Zones of Bolivia, Peru, Colombia and Venezuela that are climatically similar to two Bolivian high land wheat production sites for the 5-month optimal crop growth period (+20% similarity for precipitation and evapotranspiration, +10% similarity for maximum and minimum temperature).

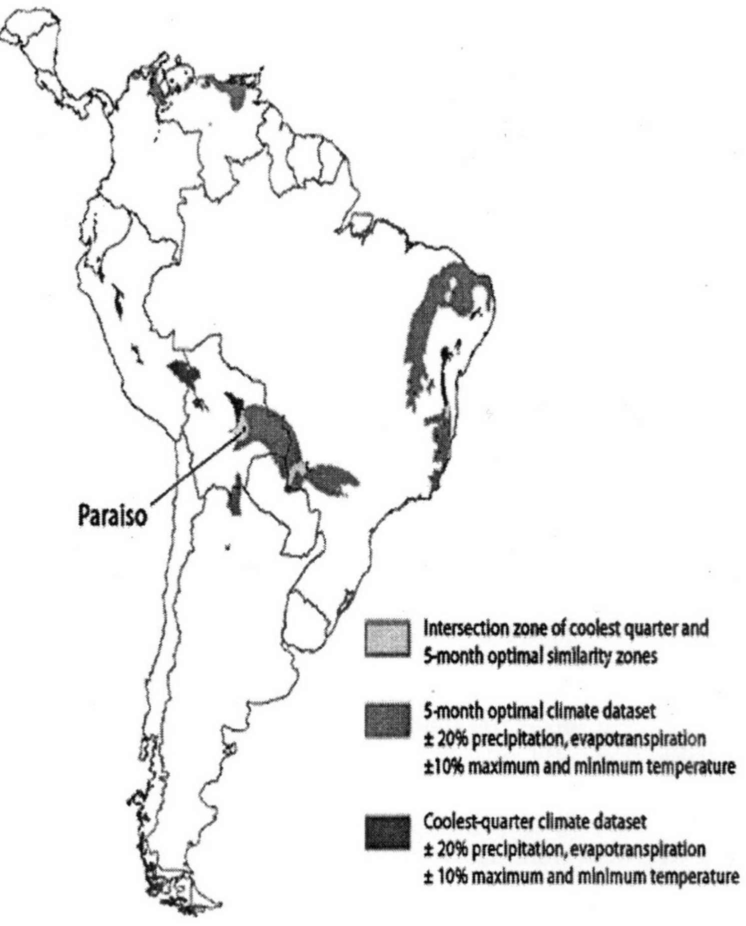

Fig. 2.4: Zones that are climatically similar to the lowland wheat production site, Paraiso, Bolivia, for the coolest quarter of the year, the 5-month optimal crop growth period, and the intersection zone of both.

In a different application and on a different scale, farmer experimenters in the Mixteca region of southern Mexico, one of the country's poorest areas, used site similarity analysis to identify locations elsewhere in Mexico with climate and soil conditions similar to their own (*See fig. 2.5 on page 25*). This information was then used to plan a study tour of research and farmer experimentation in these similarity areas. The farmers returned home with several ideas that they have begun to test, among them the use of crop residue mulches and drip irrigation for fruits and vegetables (J.C. Velásquez, 2000, unpublished manuscript).

INTERFACING MODELS WITH GIS

More complex comparisons may be beneficial to address some situations. Stakeholders may want to examine trade-offs for different scenarios. For example, are the productivity gains from conservation tillage likely to entirely offset the value of crop residues for animal feed? How will system performance vary over time, particularly in extremely dry or wet years?

Process-based simulation models can "grow" a virtual cropping system over many seasons, quickly and inexpensively. The output, in effect, extends the reach of science beyond the practicable time horizons of most research programmes, while making it possible to examine variables that are difficult or costly to monitor at the field level (e.g., nitrogen leaching and volatilisation). By interfacing GIS with simulation models, researchers can develop simulated performance surfaces that portray the likely biophysical consequences of a technology over space and over time (Hartkamp et al. 1999).

Using this approach, CIMMYT scientists examined how the performance of conservation tillage with residue retention might vary over space and time in western Mexico (Hartkamp 2000). Two key factors in the simulations were weather and soil type. Through collaboration with the International Fertiliser Development Center, a residue retention module was added to the Decision Support System for Agrotechnology Transfer suite of crop models (Tsuji et al. 1994).

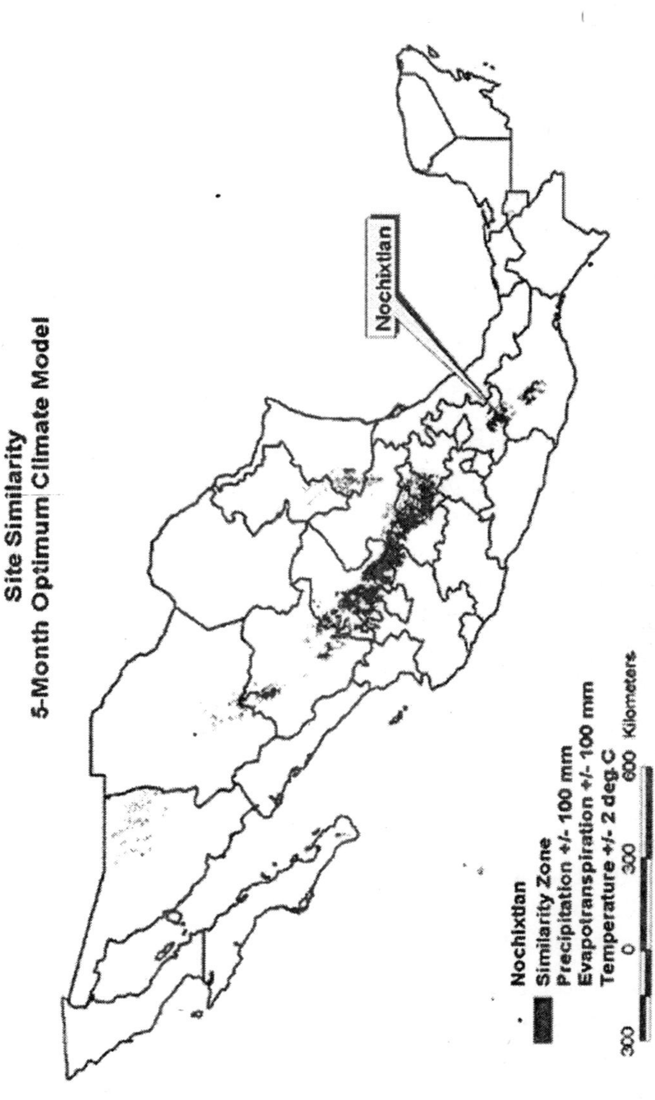

Fig. 2.5: Areas throughout Mexico that posses climate and soil conditions similar to those of Nochixtlan, a village in the Mixteca region.

Outputs from simulations (*See fig. 2.6 on page 27*) were compared with experimental results and with the researchers' own experiences. Using the resulting maps, researchers and decision makers were able to assess the simulated effects of conservation tillage on run-off and erosion, organic matter, soil structure, and moisture conservation. For each soil type, maps produced using the simulations show differences across the region in the biophysical performance of the practice. Impacts can be expressed in terms of various factors, including yield, stability, biomass, and the organic carbon and nitrogen use efficiency of the soil. The methodology thus shows promise for providing information for a range of stakeholders; the maps and other outputs can help NGOs, farmer groups, and researchers determine where conservation tillage may be most appropriate for farmer experimentation and adaptation.

LAND TYPE AND FARMER CATEGORIES

The GIS-based applications described above emphasise regional, national, or international comparisons that can be used to guide scaling out. However, more modest tools and methods that feature comparisons across farms can serve the same purpose. It has long been known that many farmers recognise different land types within a farm. These land types frequently follow the toposequence. When problems, causes, and intervention points are specific to land types, scaling out activities can be guided by the typology. When land types are replicated across large areas of the landscape, efficiencies in scaling out can be considerable. Often, of course, farmers with different resource endowments use different management practices for the same land type. Consequently, measures to foster scaling out must also consider farmer categories and cross-land type interactions within farms.

In the rice-wheat systems of the Indo-gangetic Plains, rainfall, water control, and soil texture tend to follow an east-west gradient. However, water control and soil texture in specific locations are also influenced by land type, i.e., lower, middle, and upper terraces, within a toposequence. Even though a

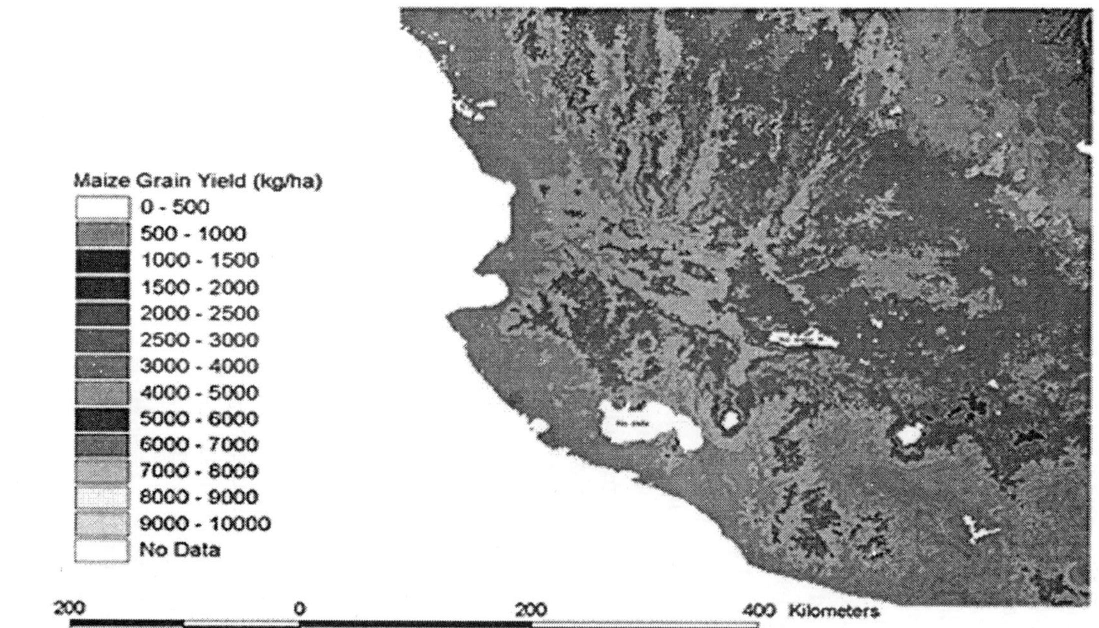

Fig. 2.6: Simulated 12-yr., average for maize grain yield under a conservation tillage system (maize-fallow with 33% residue retention) in the state of Jalisco, western Mexico, produced using Decision Support System for Agrotechnology Transfer models linked to geographic information system databases.

land type may be known by different local names in different parts of the Indo-gangetic Plains, its characteristics, uses, and management are often similar (Harrington et al. 1993). Lower terraces are characterised by heavier soils and relatively poor drainage and are more likely to be devoted to long-duration, traditional rice cultivars. Middle terraces have somewhat lighter soils and fewer drainage problems and are typically sown to modern rice and wheat varieties, at times mixed with other crops. Upper terraces have the lightest soils of all and tend to have greater agroecosystem species diversity. Here rice and wheat are sown, as well as pigeonpea, sugarcane, and vegetables.

The usual problems of rice-wheat rotations in the Indo-gangetic Plains, in particular, late sowing, high costs for tillage and establishment, low water and nutrient use efficiency, soil fertility decline, reduced agroecosystem species diversity, salinity and sodicity, waterlogging, and excessive water pumping leading to groundwater depletion, unfold differently in each land type. Similarly, intervention points change across land types but also by farmer category.

To give one simple example, it has become clear that the establishment of wheat after rice is best performed by inverted-T, zero-till seed drills drawn by four-wheel tractors for larger-scale farmers and on upper terraces. However, for smaller-scale farmers on middle and lower terraces, wheat establishment typically is best performed by surface seeding (Hobbs et al. 1998). In this practice, pre-soaked, pre-germinated wheat seed is broadcast into a standing rice crop as water is being drained off. The pre-soak is a manure slurry that makes the seed unappetising to birds. If the timing is right, soil moisture substitutes for tillage in reducing soil strength, so that roots follow the water down the profile. In both zero-till and surface seeding, there is considerable room for farmer testing and local adaptation.

In another example, farmers in southern Zimbabwe distinguish among "vlei" bottoms (wetter areas where rainfall accumulates through natural drainage), homestead gardens (with soils that benefit from crop residues, leaf litter, household waste, and farmyard manure), and toplands (with soils of low fertility and low water-holding capacity, relatively distant from the household). These different land types are managed very differently with respect to crop selection and rotations, the application of organic and inorganic fertilizers, soil fertility management, and so on (Z. Shamudzarira and C. Vaughan, 2000, unpublished manuscript). In addition, farmers with many draft animals manage land types differently from farmers with few draft animals. Nevertheless, these land types and farmer categories are replicated across much of southern Zimbabwe and adjoining areas of South Africa and Mozambique.

Exciting practices for addressing important problems, once characterised in terms of land type and farmer category, can be shared widely with farmer groups, NGOs, researchers, and other stakeholders in areas where these same land types and farmer categories prevail. Once again, the intent is to make the exciting practices available as new options to be mixed into local learning processes, not just for "cookie-cutter replication."

3

Research and Technical Innovation in Modern Agriculture

PUBLIC DISILLUSIONMENT WITH AGRICULTURAL RESEARCH

Compared to the Nobel prize-winning heights of the Green Revolution, the status and credibility of conventional or mainstream agricultural science have fallen drastically over the past decade, to the point that many people now see agriculture as a threat to the environment and to human health. The health concerns that the general public associates with agriculture fall into several areas:

- failures of modern agricultural science, such as the BSE (mad cow) disease syndrome and the foot-and-mouth outbreaks in the United Kingdom;
- bioengineered crops that introduce new allergens into the food chain, as occurred with Starlink corn, which contained a protein whose characteristics were associated with allergens and which was not approved for human consumption but nevertheless entered the human food chain;
- threats to the environment by genetically improved animals and fish such as salmon, which, were they to

escape from their commercial pens, have a mating advantage that would threaten the genetic integrity of wild stocks;

- the possible toxicity of genetically engineered crops to other life forms, e.g., BT corn, which could decimate Monarch butterfly populations;
- the development of herbicide resistance in weeds, e.g., generated by genetically engineered, herbicide-resistant soybeans that require the application of only a single herbicide.

These concerns are not restricted to high-income countries. For example, the large-scale spread of integrated pest management among thousands of small farmers in the tropics, where pesticide use has been greatly reduced, is motivated by their experience of the serious ecological imbalances and human health problems that pesticides can cause (Pingali and Roger 1995, Thrupp 1996). The way food is produced, the role of science in food production, the effects of new food production technologies on human health, and the well-being of ecosystems have become major political issues and topics for headline news. At local, regional, and global levels, the capacity of ecosystems to support human consumption of food and environmental goods and services is seen as threatened by modern agricultural practice.

Advances in agricultural science have the potential to significantly increase production in response to the globalisation of trade and the food needs of a burgeoning world population. However, many perceive these advances as sources of undesirable changes in land use, climate, biodiversity, and the evolution of new diseases. As a result, a number of countries have based their national agricultural strategies on the World Conservation Strategy, which has an explicit priority requirement to " ... reduce excessive yields to sustainable levels ... " (Adams 1990:44).

Concern for the negative effects of agriculture on natural capital stocks is both an expression of the conservation ethic and a response to international poverty, famine, and disaster.

This critique of mainstream agricultural research in the public sector is also applicable in societies in which national science bureaucracies are weak and research has been dominated by international entities financed by overseas development aid. The capacity of poor countries to withstand external shocks such as war, climate variations, and indebtedness depends partly on their natural capital. The diversity of natural capital gives it an important advantage over man-made capital in enhancing the survival of poor farmers in low-income countries, given that diverse ecosystems are more resilient to shocks and stress (Conway 1985, 1987, Pearce et al. 1990).

As a result of this critique of mainstream agricultural science, political support for mainstream agricultural research in the public sector is in decline (Andersen 1998), whereas increasing amounts of resources are now being devoted to environmental conservation and disaster relief for low-income countries. This loss of support reflects the disenchantment with modern agriculture in high-income countries and with research and advanced agricultural technology in both high-and low-income countries. Agricultural research is not unique in this loss of credibility: practitioners of rural development have also gone through a process of being proved consistently wrong (Chambers 1997). Critics of mainstream agricultural science find the research establishment "...incapable of delivering social equity, economic efficiency, and ecological integrity in response to the decline of rural society and deepening crises in the depletion and degradation of water, soils, flora and fauna..." (Campbell 1998:232-233). The rates of return on investment in agriculture in developing low-income countries are very disappointing. Much evidence is cited to support the view that returns on agricultural development projects have been lower overall than in other sectors such as health or education; in addition, the capacity of agricultural development projects to sustain momentum after external donors withdraw is seen as relatively low (Pretty 1995). Proponents of institutional change to support the development of sustainable agriculture find it hard to identify a clear role for agricultural research in this process. Röling and Jiggins (1998:297) state that "...the old role of developing technologies FOR farmers seems to clash with the logic of ecologically sound farming, while a new role [for research]...seems not to have clearly emerged."

SHORT-TERM SUCCESS vs. LONG-TERM FAILURE

For many years, mainstream agricultural science has been committed to a reductionist treatment of agricultural productivity and agriculture's role in development, as if these can be managed separately from natural resource management and ecosystem health. For example, one analysis of mainstream agricultural science equates natural resource management with location-specific adaptive research of little strategic value that is unlikely to produce internationally useful public goods and unworthy of significant levels of public sector investment (Anderson 1998).

A different type of analysis shows that, although discounting externalities may be a rational, albeit shortsighted, outcome of managing food production to maximise productivity and provide the lowest possible prices to consumers, it has significant costs in terms of human and ecosystem health (Pretty et al. 2000). Externalities related to agriculture affect the production and consumption behaviour of individuals, households, interest groups, business firms, and governments. When producers shift some of the costs of production, whether health or environmental, onto other stakeholder groups, it becomes possible for them to produce at a higher level of output and profitability than if they had to offset the full costs against the gains. An important consequence of this is that, when negative externalities are present and costs shift away from producers onto others, the output that producers consider optimal is actually higher than the output deemed optimal by society as a whole.

A choice has been made to increase production and keep food prices low rather than provide a transparent accounting of the costs of care for human and agroecosystem health in agriculture (Pretty et al. 2000). With this in mind, conventional agricultural research has focussed on the short-term goals of increasing yields and economic returns to producers and on delivering relatively low-priced food to consumers. Good historical reasons explain this focus, and extensive critiques, justification, and refutations of it abound, and will not be repeated here.

The main point is that agricultural research that focuses on increasing productivity and ignores externalities is largely built on technologies that maximise biological uniformity and sidestep, minimise, control, or destroy the natural biological processes that are essential to the stability and resilience of natural ecosystems (Folke et al. 1998). This is an issue for farmers in low-income as well as high-income countries. For example, small farmers in low-income countries are demonstrating how food production can be increased by enhancing the biological productivity of the soil, microenvironments, and even whole catchments (Pretty and Hines 2001). Ironically, perhaps the most powerful interest groups driving popular disaffection with the reductionist approach away from biological uniformity and toward diversity are well-off consumers who have the luxury of choice and whose preferences are expressed via specialised global markets. Their heightened concerns about the health and environmental implications of agriculture are creating significant demands for organic farming and niche products in subsegmented markets, and are encouraging farmers to diversify and integrate their activities with natural resource management (C. Wheatley, personal communication).

Research that treats environmental resources such as land, water, and biodiversity as discrete factors of production has, in fact, succeeded in increasing productivity in the short term. From this perspective, it is possible to maximise the productivity of individual factors of production by focussing research on improving management efficiency with new technologies. Because it works against, rather than building on, natural diversity, mainstream agricultural science is identified with processes that undermine ecosystem capacity to withstand or recover from shocks and stress. Increased production based on reducing variability slowly changes the functioning and resilience of the ecosystem that sustains the production of food and of the resources on which it depends. If variability and disturbances are not allowed to enter the system, they accumulate and return at a later stage on a much broader scale. Diminishing variability tends to increase the potential for larger-scale, less predictable, and less manageable disturbances that can have devastating effects on ecosystems (Ludwig et al. 1997); it also reduces the capacity of the ecosystem to provide the environmental services that are essential to material and energy stocks and to the flows that are important for food production. Agriculture based on blocking out variability or "disturbances" that are endogenous to the cyclic processes of ecosystem development is an expression of the rational fool syndrome, and the reductionist treatment of natural variability is the "...short-term success that leads to long-run failure..." (Folke et al. 1998:458).

Integrated approaches to research on agriculture and resource management have to bring about three critical changes to achieve the paradigm shift needed to lever food production and ecosystem and human health out of the current unsustainable syndrome. First, research for agriculture must be understood as one way of managing natural resources to support human livelihood. As such, it must build a strategic research agenda that is based on working with variability and view functional ecosystem disturbances as desirable features of sound resource management and food production. Second, research needs to be linked to active processes of adaptive management in ecosystems of high priority to important stakeholders engaged in both research and management. Third, the appropriate organisations for carrying out this research need the support of the emergent interest groups and learning communities that are propelling a paradigm shift outside of and virtually despite mainstream agricultural science.

AN EXAMPLE OF WORKING WITH VARIABILITY

A classic example of a paradigm shift lies in the history of the management of African pastoral systems (Ellis and Swift 1988). Recommended methods of reducing overgrasing in these pastoral systems included group ranches, grazing blocks, and associations in which pastoralists were confined to particular tracts of land to better regulate the interaction between animals and plants and raise productivity. Over time, these new management methods were found to destabilise grazing systems that are characterised by intra-annual variability resulting from frequent drought. In contrast, pastoralists using traditional methods cope with multiyear drought by dispersing into small herds and groups over a wider area, thus expanding the spatial scale of exploitation. In non-drought periods, pastoralists ensure that unused space or an ungrased reserve is available for periods of drought by stocking some areas in the ecosystem well below their average carrying capacity (undergrasing) while overgrazing others. This stabilising mechanism relies on mobility, whereas the modern management strategy is based on confinement. In other words, recommendations that do not factor in variability and disturbance in the ecosystem often lead to long-term failure. Research had to define alternatives to conventional management of grazing systems that functioned at the ecosystem level, took into account hierarchies of interdependent subsystems, and were effective over the long term. Technical packages designed for a reduced spatial scale and short time horizon could not cope with the variability in the system, and indeed became associated with increased degradation in the long run (Ellis and Swift 1988).

An example of working with adaptive management.

The experience of the Forages for Smallholders Project (FSP) (P. Kerridge et al., unpublished manuscript) with the introduction of legumes and grasses into small-farm production systems in southeast Asia illustrates how a mixture of participatory learning and research can support an integrated approach to food production and natural resource management. This type of approach fosters adaptive management that is responsive to and supportive of variability in ecosystems.

The project is introducing of a large variety of optional, new, multipurpose legume and grass species using participatory methods that enable farmers to learn about diversity and devise a variety of niches for different species in their systems. The use of a menu of diverse options for participatory learning is fundamental to this approach. Farmers learn, invent, and validate new ways of using diverse species, and the ways they choose are becoming more complex as the farmers gain experience.

Over time, farmer preferences for varieties and characteristics have changed. At first, animal feed, contour hedgerows, and intensive cut-and-carry plots emerged as the main ways in which farmers integrated several species into their agricultural systems. Subsequently, other ways of integrating legumes and grasses into living fences, ground covers, and small grazed plots evolved for many different end uses. Species and genetic diversity as well as new ways of coupling plants and animals in the system, e.g., growing shrub legumes for forage as part of contour rows for soil erosion control, have been developed as a result of farmer innovations. Participatory research fostered adaptive management that enabled the project to encompass the diversity and complexity of entire agricultural systems. The researchers, whose initial ideas about appropriate options for each farming system did not agree with those of the farmers, eventually revised their opinions to reflect the fact that the farmers' grasp of complexity was different than their own.

Researchers' innovations have also been adapted in unexpected ways. For example, on steeply sloping lands, farmers introduced contour plantings of Paspalum atratum, which establishes much more rapidly and cheaply than the heavily promoted vetiver grass and can also be used for fodder. P. atratum competes with associated crops, but not to the same extent as other types of barriers. The farmers took an integrated approach to managing small areas planted with legumes for forage and also used alternative feed sources from adjacent forest and waste areas. In this way, they highlighted for the project the potential of managing both types of land use for the preservation and increased use of indigenous species.

NEW ORGANISATIONAL CONCEPTS

Eighteen projects in geographically representative regions of the United States make up the Integrated Farming Systems (IFS). These projects are linked in a network that makes it possible to share knowledge and experience among farmers, scientists, consumers, policy makers, bankers, and producers and distributors of agricultural inputs. The goals of the IFS are to help farmers discover and use more sustainable farming practices and to help communities identify and overcome barriers to adopting more sustainable practices and systems (Fisk et al. 1998). A highly important feature of the network is that it also seeks and creates leverage for catalyzing policy change on the basis of experience gained in the context of an adaptive management approach to sustainable agriculture. The use of formative evaluation to document progress, reflect on what works, and determine the next steps to take has been crucial to this catalysis (Fisk et al. 1998).

As stated above, one goal was to help farmers find and use more sustainable farming practices with the support of their communities and of policy makers. To achieve this, the IFS set out to establish a "learning community" that practices the five disciplines identified by Senge (1990) as the requisites of a learning organisation. By self-consciously working together around the idea of creating a learning community, the IFS has successfully built a support structure for farmers and community spokespersons that favours sustainable farming. Farmer-driven research and educational organisation are key elements: farmers conduct on-farm research, educate other farmers and community members, and provide laboratories that foster leadership. As a result, a cadre of community-based spokespersons has been developed who can effectively carry messages about their needs and achievements to decision-makers and opinion leaders in their own communities and centers of policy activity. Policy makers in the United States are looking to community-based organisations for information about the effects of policy changes at the farm and community levels. Among its 18 projects, the IFS has provided some powerful demonstrations that have helped to educate policy makers (Fisk et al. 1998).

An important ingredient of this success is an effective strategy for working across different organisational scales. To achieve this, the IFS helps communities to engage large institutions by finding the constituency that these organisations are most likely to listen to and building a relationship with that constituency. The IFS has been highly effective in communicating information and developing leadership skills to catalyse action on national policy issues. One result of this is that the U.S. Department of Agriculture now uses the IFS model.

Agriculture as a way of managing natural ecosystems for human well-being.

Several novel approaches are helping to integrate agriculture and natural resource management, although there is still no coherent unity of thought or organisation. Scientists have called for a broader focus and created new names for these approaches, all of which emphasise a balance between productivity and environmental health objectives and, to varying degrees, view agriculture as one way of managing natural ecosystems for human well-being. Much of this thinking parallels the processes that have been developing for many years within applied research projects and programmes focussed on adaptive farming systems (D. Gibbon, unpublished manuscript).

These new perspectives are built on the premise that human health and livelihood depend on the functioning of ecosystems that generate essential natural resources and environmental services, play a fundamental role in supporting life, and are essential to the material cycles important for food production. Environmental services include regulating climate variability, water quality, and water quantity; controlling floods and pests; assimilating wastes; recycling nutrients; generating soil; pollinating crops; and maintaining the composition of the atmosphere, biodiversity, and the landscape (Conway 1997).

The idea that the methods used by traditional cultures to manage food production are integral to natural resource management has also helped to break down the reductionist separation of agriculture and ecosystem management. Many documented examples illustrate indigenous systems for managing whole landscapes as integrated farming systems; these range from the upland forest all the way to the coral reef (Folke et al. 1998).

Once agricultural systems are understood to be parts of, and managed like, ecosystems, research needs to take into account the effects of agriculture on the environment. Ecosystems are "holarchies" made up of many different components or "holons" that operate together. A holon is a whole system that is made up of smaller parts and at the same time forms part of a larger system (Allan and Starr 1982, Giampietro and Pastore 1999). This concept is useful in illustrating how agriculture involves interventions in component systems that are hierarchically organised on several scales: soil microorganisms, populations of species, fields in a landscape, etc. In contrast to mainstream agricultural science, new approaches that integrate food production and environmental health objectives analyze agriculture as a set of interventions in a hierarchical system. This helps to emphasise the fact that ecologically sound conditions for growing healthy animals and plants must usually be created at more than one system level, that is, higher than the plot or even the farm, and must be managed on a landscape scale (Holling et al. 1995). This was the approach used to define a successful management strategy for African pastoral systems.

Agricultural systems manifest an important characteristric of holarchies or hierarchies of interdependent subsystems: what is good in the short term is not necessarily good in the long term (Giampetro and Pastore 1999). Understanding the holarchic structure of agricultural systems is crucial to the recognition that knowledge of what goes on at one level of a holarchy does not necessarily provide sound information about what is going on at other levels. For example, increasing productivity at the level of individual plots can reduce the overall productivity of a whole landscape such as a watershed, and pest and disease control to improve plant health in the farming system can jeopardize human health in the larger production-to-consumption system.

Research on agriculture treated as one component of human intervention in complex ecosystems will need to be increasingly strategic in nature. This type of research must deal with a high level of uncertainty based on patterns that are repeated in an irregular way. These are also known as fractals: generic patterns repeated in a self-similar way at many different levels that are always recognisable but never exactly the same. In principle, research can design an agricultural system or subsystem with a generic pattern and potential, establish the conditions required for it, and predict the dynamics that will produce a recognisable variant of this system. Action research is an approach that has potential applications in this kind of situation because it allows researchers to intervene in an agroecosystem, study the effects of the intervention, and apply this understanding to a second intervention. Research efforts of this type can be conducted to push a system into a phase of instability and change within certain constraints, monitor the resulting adaptive management process, and use the feedback to develop management principles. However, research cannot determine the specific form of the generic pattern that may emerge. For this reason, research that builds on variability will not produce technological "silver bullets" in the way that reductionist agriculture has: its focus by definition must be on the strategic and the generic, on "how" rather than "what" to change in agriculture. The adaptive management of species diversity in the Forages for Smallholders project is an example of this shift in approach.

In agriculture, new technologies based on the active reduction of variability in the flow of a resource or a natural capital have, in fact, succeeded in increasing productivity at one level of the holarchy or in one of a series of interdependent systems. In mainstream agricultural science, new technology is designed to reduce and overcome fluctuations and variability in nature and to block out the natural disturbances that are functional and necessary for sustaining ecosystem resilience (Folke et al. 1998). To a very large extent, the anomalies that are fueling public disenchantment with mainstream agricultural science are instances in which success in overcoming variability in the short term in one part of the system has led to failure in the long term at another level.

ORGANISING FOR INTEGRATED RESEARCH APPROACHES

Once it is understood to be a form of natural resource management, agricultural science must definitely be subject to the value-driven preferences of diverse interest groups or learning communities. Resource management involves negotiating goals and acceptable trade-offs among multiple stakeholders, including the different learning communities. For poor farmers in semi-subsistence agriculture, there are trade-offs between satisfying the family's daily food and income needs and maintaining the viability of the natural resources required to produce them. For better-off farmers in commerical agriculture, there are trade-offs between cutting costs to capture slender profit margins and long-run investment in the management practices and technology needed to sustain productivity. Research aimed at improving the management of ecosystems for food production and other purposes has to incorporate the management objectives of the different stakeholders with regard to how best to use natural processes, cope with disturbances, and internalise externalities.

This has several implications for the organisation of research intended to integrate the objectives of food production and environmental health. The democratisation of science, which in this case is understood to mean the inclusion of lay expertise and values in the process of planning and carrying out research, may be a prerequisite to improving the quality of scientific understanding and obtaining support for the research. In particular, extended peer communities, especially the stakeholders affected by an environmental problem, need to be involved in science when cause-and-effect relationships are unclear and when research measurement and ethical aspects have a high level of uncertainty (Funtowicz and Ravtz 1993, Irwin 1995). The present debate about integrated approaches heightens the need to include lay expertise and to cease restricting the conduct of research to technical specialists. Including citizen expertise when reforming science bureaucracies is essential, because research that integrates the objectives of food production and environmental health will require input from various realms of knowledge. New kinds of organisations that aim to carry out research based on integrated approaches must be prepared to consider knowledge that has not been scientifically generated, deal with a plurality of knowledge forms rather than a unitary consensus, take into account stakeholders' concerns, and be extremely flexible. The IFS illustrates this type of institutional approach.

Institutional flexibility is required to enable diverse stakeholders to carry out a paradigm shift toward integrated approaches. Research organisations must be learning organisations that can not only conduct strategic research but also monitor and provide feedback on adaptive management projects for agroecosystems. In this respect, there is a notable lack of fit between conventional management institutions and ecosystems, in particular agroecosystems that require adaptive management, and some observers see this lack of fit as a "...crisis of institutional learning..." (Folke et al. 1998).

An important organisational feature needed to implement integrated approaches is a flexible hierarchy that enables decision-makers to interact at different ecosystem levels. Researchers need to interact with the non-scientists who form part of extended learning communities. This organisational feature is missing in conventional agricultural science bureaucracies. An organisational type with this potential for flexibility and learning was identified in the mid-20th century as an "adhocracy" (Mintzberg and McHugh 1985), but it was not formally used to manage agricultural science until recently. An adhocracy is characterised by the production of complex outputs that demand sophisticated innovation by combinations of experts deployed in temporary teams to work on projects. In an adhocracy, coordination is achieved less by direct supervision, performance controls, and rules than by selective decentralisation. The power to make decisions is decentralised in uneven ways, and devolves to the level or person most likely to have the expertise needed to deal with the issue at hand.

Three elements of organisational change in an adhocracy are well suited to the current need to promote organisational change in support of integrated research approaches in agricultural science. First, management usually allows patterns of working together to emerge in a self-organising way. Second, strategies can be developed by unusual nodes in the organisation and adopted by the collectivity. Third, management has to create a climate in which a wide variety of strategies can grow, watch what comes up, and tolerate the unexpected. This approach to change is consistent with the needs of research linked to the adaptive management of ecosystems.

Many of the organisational features of the adhocracy coincide with those identified with learning organisations. Senge (1990) mentions these characteristics: a willingness to take risks and experiment, decentralised decision-making, systems for sharing learning and putting it into practice, frequent use of cross-functional work teams, opportunities to learn from experience on a daily basis, a culture of feedback and disclosure, and collective vision building. Finally, the learning organisation is closely connected to its environment, which includes legislators, regulators, clients, competitors, and communities.

The disestablishment of science bureaucracies, which have been the home of agricultural research over the past 100–150 yrs., is occurring at breakneck speed in both high- and low-income countries, largely as a result of public disenchantment with, and the subsequent loss of political support for, the reductionist type of agriculture described earlier. The design of new organisational forms for carrying out alternative integrated approaches to research is also under way in a highly decentralised and self-organising process. Participatory research approaches provide alternative ways to democratise science. Social and institutional issues of strategic importance to the future of ecosystems and human health care have only recently begun to emerge as a research topic (Folke et al. 1998). Although a systematic overview of this process of self-organising change is lacking, its main features, which can be determined from the projects already in place, correspond to the principles embodied in flexible learning organisations.

Keystone properties of research that integrates food production and environmental health objectives.

Emergent alternative approaches have in common an integrated approach, which treats agriculture as one element of human intervention in ecosystems. Although numerous and diverse, these integrated approaches manifest several common elements; in particular, they:

- integrate research and adaptive management of food production with ecosystem and human health by taking ecosystem and human health externalities into account in food production and marketing;
- recognise that agricultural science is not the exclusive domain of scientists and that internalizing environmental and human health externalities involves negotiating trade-offs that depend on value judgements by different stakeholders;
- conduct strategic long-term research that works with and for sustained ecological, economic, and socio-cultural variability;
- assign priorities to strategic research based on an understanding of the dynamics of cause and effect in ecosystems (including agroecosystems) across hierarchical system levels;
- approach agriculture as an element in ecosystem management, from an area perspective, and not exclusively as commodity-based or as factor research;
- incorporate adaptive management into research so that these domains interact to provide rapid feedback to the relevant learning community responsible for food production and for ecosystem and human health;
- carry out adaptive management experiments that include interventions of different types (e.g., policy, technology, collective action) at various scales to achieve understanding when scale and uncertainty make positivistic experimentation difficult;
- support the development of information management decision support and other ways of being responsive to

and supportive of ecosystem variability as premium strategic research products (e.g., good monitoring and diagnostic tools able to detect change);

- ensure that their scientific method and research organisation are diverse and responsive to variability and built on the principles of learning organisations, so that both provide capacity for enhancing the resilience of ecosystems;
- organise decision-making hierarchies that involve scientists, managers, and other kinds of stakeholders in research and adaptive management that match the ecological hierarchy of scales affected by the pertinent stakeholder objectives and ecosystem management issues;
- incorporate into agricultural research explicit considerations of the preferences of competing interest groups in different kinds of trade-offs between productivity and environmental health;
- foster learning communities with research and adaptive management functions, ensure interaction between research and management expertise, and break down barriers between different kinds of knowledge (e.g., expert and layman); and
- organise learning communities to work with multiple stakeholders in areas such as negotiation platforms, conflict resolution, facilitation, participatory research, and learning to make the value judgements required to arrive at acceptable trade-offs among competing goals for the same resources.

4

Robot Integration in Agro-Technology

ROBOT INTEGRATION

Robot integration involves combining machine vision with mechanical motion to provide automated systems. This can have many different applications, such as in agriculture and the automotive industry, with various levels of sophistication, from vision assembly and conveyor tracking, where vision simply aids the production line, to 3D and small-lot assembly, with tightly integrated, 'near-human' vision.

The simplest of these would be a single-axis robot with a single area mount camera and no conveyor tracking. This kind of system could be applied in medicine and the automotive industry, and would be simple pick-and-place mechanisms, small-part palletising, and screw driving. Next up would be a system involving a four-axis robot, still with a single camera, with simple conveyor tracking and a vibration feeder. This would be used in electronics and telecommunications, and possible applications include tracking pick-and-place mechanisms, flexible feeding and precision assembly.

The more complex applications have greater precision, and use multiple area-mount and line scan cameras (with possibly one mounted to a six-axis robot), together with multiple high-speed conveyors and multiple mechanisms. Due to the greater level of precision, these are suitable for use in food and pharmaceutical packaging applications, flexible part manufacturing, and multi-product assembly. They can give greater speed, which is necessary when producing lower-value parts.

There are many difficulties when developing such schemes – as Jan-Philippe de Broeck, of Adept Technology, explains: 'Among them is the transformation of the vision coordinates to 3D world coordinates that the robot can use to pick parts, as well as the requirement for communication efficiency and synchronisation between the vision system and the robot. The physical link is important, but most critical is the software that must be able to reliably send multiple part locations (and vision inspection results) to the robot controller in a very short amount of time to enable the highest throughput. Information needs to be sent from the vision system to the robot controller so that appropriate movement instructions can be carried out in time and with great accuracy.'

THE AUTOMOTIVE INDUSTRY

The automotive industry is one of the key areas to benefit from applications robot integration. Bin picking is one such example, in which a 3D vision-guided robot uses a camera to identify randomly placed parts in a pick-up area. A robot then picks up the part, and can transfer it to the next step on the production line. According to Adil Shafi of Shafi Inc, who has helped develop such a system, this level of sophistication 'is the holy grail of robotics'.

Processes like bin picking are so precise, with accuracy up to 0.1mm, that they practically eliminate the need for humans on the factory floor. Humans are the slowest part in manufacturing procedure, so vision-guided robotics would speed up production significantly.

Inflating labour costs also add a key incentive. As Shafi claims: 'In America, the average labour costs are $20 an hour. Using vision and robotics, the cost is $3 to $5 an hour. This is very attractive in countries with high labour costs. It allows them to keep manufacturing in their own country, while still being able to compete in the world market. What's more, with robots, there are no costs of worker injuries and there are health benefits.'

Ford uses vision-guided robots for high-precision tasks that are impossible for humans. For example, cylinder heads are highly sensitive and must be kept immaculately clean. In the past, high-cost precision-shipping containers had to be used while fitting the heads to the blocks. Now, precision automation makes this possible with much cheaper containers that can hold the parts in less precise positions. Ford also uses vision-automated vehicles to transport parts, and load and unload racks.

THE USE OF VISION-GUIDED ROBOTICS

The use of vision-guided robotics is not confined to the automotive industry – you can even see an advanced application of robot integration in practice in the rose gardens of The Netherlands. Agrotechnology and Food Innovations is currently testing a fully automatic robotic rose harvester. The machine includes three robots and five cameras. It makes use of both 2D and 3D vision to locate the roses from eight-metre-long gutters, and to decide where would be best to cut the rose before actually doing the job. The third robot then retrieves the cut rose.

Robot and vision integration is not just limited to vision-guided motion. Aptura has developed a new method of quality control of sand cores in the engine block casting process, in which a robot transports the parts and presents them to a number of cameras that capture images of the part and check for abnormalities. According to David Dechow, of Aptúra Machine Vision Solutions: 'The advantages of this are that it is fully automated, with limited human intervention. It does not necessarily reduce labour cost, but it does increase productivity, as it ensures error detection. If an error were to remain undetected the whole thing would have to be scrapped.'

Vision servoing is another technique, and is currently at the cutting edge of robot integration. 'The idea is wonderful,' says Dechow. 'A camera watches and guides a robot to a moving object, without encoders or any other sensors.' Although it is hoped that it will have many applications in the near future, it has not fully matured as a technology and, as such, is not widely in use yet.

Valerie Bolhouse, automation systems specialist at Ford, agrees. 'Vision servoing would be the biggest thing yet to come, and we would see a lot of assembly applications opening up. At the moment it is very expensive to accelerate, then stop, then accelerate the assembly line while the heavy parts are fitted.'

Although the development of such schemes has been blooming in recent years, this technology hasn't always been accepted by the machine vision industry. 'The development, and reliability has increased dramatically in the last few years,' says Shafi. 'The '90s were very disappointing. However, the internet has increased development by giving factory workers computer literacy, so training and support are much easier.'

Financial concerns are obviously a big factor. Currently, it has been estimated that 18 per cent of production costs come from integration, making it the largest proportion. However, this is not set to always be the case. According to an AIA market study, the trend is for the average system price to decrease rapidly with time, meaning robot integration can become more affordable for smaller companies. With labour costs growing all the time, it is increasingly an appealing option.

According to David Dechow, the increase in computing power, and the development of smart camera architecture, which cut the components' costs, has been a big factor in its success. 'And of course, successful applications created a growing recognition that use of machine vision can be successful.' Smart cameras also facilitated robot integration by reducing the number of wires needed to output results, as the computer is fitted with the sensor.

Whatever the challenges, it seems inevitable that with the current trends of increased acceptance and decreased prices, vision-guided robots will have a much more prominent place on the factory floor in the near future.

5

Agro-Technology and Food Innovations

SUBSTITUTING ISOSORBIDES FOR PHTHALATES

Phthalates with short ester chains (< C9) have been suspected of negative health effects. Since September 2004, Europe has not allowed them in toys and personal care products for children. The novel plasticizers in this technology are based on isosorbide. They can substitute for phthalates as they are not suspected to be toxic and they have comparable — and sometimes better — properties. With this novel isosorbide plasticizer family, a new platform for phthalate substitutes comes into reach. The patented synthetic route is unique with 99 per cent conversion and 90-95 per cent selectivity.

The substitution of phthalates by other plasticizers is not unique. For specialty applications, other plasticizers with different properties are applied. However, except for these isosorbides, there is no "platform" chemical that can replace phtalates: for different applications, various chemicals are being used. Moreover, there are only few chemicals that have properties comparable to those of phthalates. This novel isosorbide family is unique in the sense that the family has properties comparable to phthalates and at the same time can cover the same broad range of applications:

- Compared to the disputed phthalates, isosorbides can offer a green, undisputed alternative. Current toxicity data show no acute toxicity, no sensitization, no mutagenic effects (Ames test), and no estrogenic effects.
- Moreover, the properties of the isosorbide esters can be tuned with regard to compatibility with the plastic, in migration, in plasticizing efficiency, and in volatility. For instance, the isosorbide C-8 diester compares in all properties (except suspected toxicity) with the phthalate C-8 diester.
- A major advantage in coating applications is that the novel isosorbide plasticizers are generally far more transparent. This is of importance for applications such as top coatings on floor coverings, in wallpaper, and in medical disposables.
- Isosorbide esters have favourable migration characteristics compared to phthalates as well as to adipates, which is advantageous in food-contact applications such as food wrapping and conveyor belts.
- Isosorbide esters are high-solvency plasticizers, which can compete with BBP or benzoates in applications such as adhesives, sealants, and inks.
- Because isosorbide esters contain no aromatics, they are expected to produce less smoke when burned than do phthalates.
- Isosorbide esters are high-boiling-point solvents that do not degrade at temperatures up to 250°C. In PVC, the heat stability is better than those of phthalates like BBP.

Other advantages depend on the application and vary from reduced volatility and better thermal stability, to better compatibility with PVC and reduced water absorption.

PLATFORM TECHNOLOGY FOR MICROBIAL PRODUCTION OF AROMATICS

This platform technology uses microbes to produce a diverse array of higher-functional aromatic chemicals derived from tyrosine or para-HydroxyCinnamic Acid (pHCA) used as monomers for novel polymers, adhesives, coatings, aroma chemicals and flavours, pharmaceuticals, bio-cosmetics, and health and nutrition products. Several higher-functional and expensive chemicals that are difficult to produce through conventional chemistry can be made from these lower-cost base materials. Production of these chemicals using renewable feedstocks at lower prices would expand the market for these chemicals and provide a competitive advantage.

Tyrosine, pHCA, para-hydroxystyrene, and acetoxystyrene monomer are valuable intermediates in their own right for cosmetic and pharmaceutical applications, and acetoxystyrene is polymerized and used for photoresist in electronic applications, and numerous coatings such as anti-corrosion coatings for automobiles. This technology offers a renewable resource, economic route to these compounds beginning with the fermentation of glucose.

Crafting a biological route to acetoxystyrene monomer for example, requires overcoming significant technological challenges:

- Maximize carbon recovery from glucose.
- Develop an economically productive catalyst for conversion of Tyrosine (TYR) to para-HydroxyCinnamic Acid (pHCA).
- Minimize catalytic inhibition by the produced pHCA.
- Mitigate the cell poisoning by pHCA and para-hydroxystyrene (PHS) in the bioconversion step of pHCA to pHS.

The technology is a two-step process to produce pHCA. In step one, E. coli converts glucose to L-tyrosine (TYR) in a typical fermentation process. Step two uses an engineered E. coli to convert the L-tyrosine to pHCA in a batch bio-conversion process. This second, engineered, E. coli expresses the enzyme Phenylalanine Ammonia Lyase (PAL), which has Tyrosine Ammonia Lyase (TAL) activity. The pHCA produced in this step is then isolated and purified. The technology can be implemented in existing typical glucose fermentation equipment with some modifications to isolate the tyrosine and pHCA solid products. Demonstrated productivities and titers for the two steps are within commercial bio-specialty range and attractive economics are anticipated.

EASILY RECOVERABLE, ARBITRARY PROTEINS PRODUCED BY RECOMBINANT HOST CELLS

This strain of Rhodococcus acts as expression vector for one or more foreign genes simultaneously, to express one or more recombinant proteins in the host cell. Because this strain of Rhodococcus has a higher sensitivity to lysozyme and undergoes lysis more easily than does the wild strain, recombinant protein is more easily recovered. The cells can proliferate and express proteins at temperatures lower than E. coli; thus proteins can be produced that would be inhibited by the other host's higher temperature requirements. Applications already include production of proteins that undergo a conversion reaction and that can decompose environmental pollutants.

How this technology differs?

Currently, an expression system using Escherichia coli as a host is widely used for preparing large amounts of a recombinant protein derived from a eukaryote. This is because the E. coli is readily handled.

Rhodococcus erythropolis can also be used as a host for recombinant protein production.

R. erythropolis is one of Actinobacteria capable of proliferation at temperatures ranging from 4°C to 35°C. The biggest feature of an expression system that uses this bacterium as a host is that a recombinant protein can be produced at temperatures not higher than 10°C, such as 4°C. In systems that employ other hosts including Escherichia coli, bacteria belonging to the genus Bacillus, yeasts, and Sf9 insect cells, recombinant protein production at temperatures not higher than 10°C is quite difficult. With Rhodococcus, recombinant protein production at temperatures not higher than 10°C has allowed the production of proteins difficult to produce until then, for example, proteins that inhibit the proliferation of host cells, proteins that become insoluble at temperatures around 30°C, and proteins derived from organisms adapted to low temperatures.

E. coli is a mesophile, and although it grows at 18 to 37°C, culturing temperature for expressing the recombinant protein should also be within this temperature range. However, since the protein derived from a eukaryote shows activity in the same temperature range, some proteins, when made to express in E. coli as a recombinant, may inhibit the growth of E. coli. Consequently, significant quantity of recombinant protein may not be obtained.

Besides those using E. coli, expression systems using eukaryotic cells such as Saccharomyces cerevisiae and Pichia pastoris as a host may not express the protein efficiently either unless the culturing temperature is around 30ºC, with similar difficulties for the recombinant protein.

Purpose-designed proteinic production using Rhodococcus is already under study in Japan as a method for decomposing environmental pollution by making proteins that undergo a conversion reaction.

TRACK-CONVERSION KIT FOR CONSTRUCTION VEHICLES

A common problem encountered in the construction sector is that a wheeled scraper or other heavy construction equipment becomes stuck due to unstable terrain, such as mud, sand, ice, or snow. Often a bulldozer is required to pull or push vehicles that are stuck, resulting in additional costs. Not only does the construction company have to bear the cost of the bulldozer and operator, it continues to pay for the stuck piece of equipment and its operator. It is a common scenario to wait up to 18 hours for tyre repairs if the equipment is in an inaccessible area or if the tire repair company needs to order a replacement tyre. The company will continue to incur the cost of the machine and the operator while waiting on repairs. Of greater magnitude are the costs incurred due to the delays caused in project completion.

These problems can be solved when construction vehicles are no longer at the mercy of unstable terrain conditions. Tracked vehicles are the solution to traversing terrain that is inaccessible to wheeled vehicles. A tracked vehicle eliminates the need for tyres, and enhances maneuverability. The principal design advantages of tracked versus wheeled vehicles are that tracks are in contact with a larger surface area than would generally be the case with a wheeled vehicle. As a result, they exert a much lower force per unit area on the ground being traversed than does a conventional wheeled vehicle of the same weight. This makes them suitable for use on soft, low-friction, and uneven ground such as mud, ice, and snow.

The TITM kit allows the conversion of a vehicle that customarily runs on wheels into one that runs on tracks. The TITM kit converts each wheel assembly on a construction vehicle into a track drive. The track drive in the TITM kit is operationally connected to a transmission and drive shaft of the construction vehicle. The TITM kit uses the elevated drive sprocket design, thus improving durability and ease of maintenance. The kit comprises a pair of support braces, a base plate, four shock absorbers, four axles, a pair of torque bars, and four spindles. It is specifically designed to attach to existing heavy construction vehicle, with different kits for each type of vehicle. The TITM kit enhances a driver/operator's safety by stabilising the vehicle and improving its maneuverability and decreasing vibration. It also improves traction for off-road trucks on unstable soil conditions such as sand and loose gravel or stone. Installation of the TITM KIT into existing construction vehicles reduces downtime for repairs as well as man-hours required to complete the job. TITM increases operator comfort, thus increasing work efficiency and making it easier to complete projects on time. TITM is a reliable and cost-effective alternative to rubber tyres and installs directly into existing heavy construction vehicles.

PLATFORM TECHNOLOGY FOR ELECTROMECHANICAL ONE-AND TWO-DIMENSIONAL POSITION SENSOR

This technology provides an elegant, efficient, and cost-effective way to measure absolute position in one or two dimensions using a new Point Coupled Linear Transformer (PCLT). The PCLT detects movement and position on flat or curved surfaces. It has a solid theoretical foundation and employs well-established, reliable, and cost-effective production techniques. The nature of its design provides the sensor with excellent linearity, its DC output voltage allowing it to be used as a direct drop-in replacement for other position sensors. This makes it an economically viable solution with very low switching costs. The North American market for linear displacement sensors — including encoders, linear variable differential transformers (LVDTs), laser systems, magnetostrictive, potentiometric, capacitive, inductive, and ultrasonic products — totalled $338.7 million in 2006. This provides a projected global market value exceeding $1 billion. To-date, inductive sensors (LVDT/RVDT/LVIT) have dominated the market, but they have certain limitations that can be overcome by this new technology. Namely, PCLTs can be used to measure a position on flat or curvilinear surfaces and can also be extended to 2-D measurements, whereas LVDTs are limited to 1-D position sensing. PCLT also has a 90 per cent stroke-to-length range, which is 30 per cent better than LVDTs. This could result in a 30 per cent–60 per cent weight reduction of the sensor. This technology offers strong patents and IP traceability that are available for licence to suitable sensor manufacturing companies. Considering the obvious benefits in terms of cost and improved performance, it can potentially become the new sensor standard, replacing LVDTs.

About the Technology

The PCLT technology provides a good platform for non-contact absolute position sensors. The simplicity provides a solution that is scalable both in terms of physical size and from one to two dimensions. The versatile mechanical configuration allows for positioning on flat as well as curvilinear surfaces. The winding design has a solid theoretical basis, resulting in an output signal that exhibits a linear relationship with the absolute position of the ferrous core. In addition, the winding configuration can be adapted to produce higher sensitivity in predefined sensing areas.

Well-established production techniques are used to manufacture the PCLT sensor, and simple electronics are used to excite the primary and secondary windings. No on-board microprocessor is required for signal conditioning of the DC output.

The performance parameters of the PCLT sensor compares favourably with other position sensors available on the market. These are some key parameters measured on a PCLT:

- Linearity: 0.18 per cent of full scale (32mm stroke)
- Output noise: 2 mV (peak-peak)
- Drift of 0.5mV over 12 hours
- Resolution: at least 8.3μm or 0.013 per cent FS
- Repeatability: 3.2μm or 0.005 per cent FS

Various different configurations of the PCLT sensor were designed, evaluated, and characterised in small-scale R&D projects. Theoretical analysis and lab demonstration models were constructed as part of various post-graduate research projects. Intellectual property protection of this technology via patenting has passed through the PCT stage, receiving a favourable search report, and the patenting process has now progressed to National Phase. A US patent was granted in 2008, and this application claims priority to and is a continuation of an international PCT application filed in 2004, which claims benefit from and is a Paris Convention filing of U.S. a Provisional Patent Application filed in 2003. A South African Patent was granted in 2006. The European patent application is currently in the examination phase and one response to an Official Action has been submitted.

ENVIRONMENT FRIENDLY PERFLUOROALKYLS IMPART NON-STICK PROPERTIES AND RESISTANCE TO OIL, GREASE, SOLVENTS, OR WATER

This highly flexible C6 perfluoroalkyl formulation overcomes the environmental issues inherent in other perfluoroalkyl tails, and imparts non-stick properties and resistance to oil, grease, solvents, or water that are similar to

C8 fluoroalkyl tails. No other commercial chemical contains a similar structure. Chemicals containing perfluoroalkyl tails are used extensively for stain, water, oil, and grease resistance, and for their surfactant and non-stick properties. However, most perfluoroalkyl tails have come under environmental pressure because of their potential to form PFOA (PerFluoro Octanoic Acids or salts). Both Canada and the US have ceased approving new chemicals that contain C8 or longer perfluorocarbon tails in their structures. Hercules C6 products have clearances from multiple regulatory agencies internationally. In a commercially available formulation used on a cellulose substrate, it resists oil and water penetration under either aggressive (hot polar fats) or non-aggressive conditions (ambient temperatures, non-polar oils). It is already approved for paper treatment where the paper may be in contact with food. The flexible C6-tail formulations can easily be adjusted to change the molecular structures that control oil repellency, reactivity and charge, molecular weight, and water stability. Applications include textiles (stain repellency and oil resistance), coatings (aids to flow and spread, resistance to stains and deposits), and as lubricants and release agents.

Fluorocarbon Tails

Chemicals containing perfluoroalkyl (CF2)nsections, or tails, have been used in industry to provide various oil-, water-, and stain-resistant properties to a range of substrates, such as paper. However, chemicals containing perfluoroalkyl tails where n = 8 or greater have come under increasing environmental pressure. Several governments are moving to ban chemicals that contain C8 or larger tails, and Environment Canada and the EPA in the US have stopped approving such chemicals.

The Hercules formulation of perfluoroalkyls with C6 tails offers properties similar to chemicals having C8 tails, but with a much more environment friendly formulation. The molecule is highly flexible, and can be modified quickly to alter and control oil repellency, reactivity and charge, molecular weight, and water stability.

By anchoring fluorocarbon tails to a substrate (such as cellulose), where they fix, Hercules C6 molecules can achieve oil/fat/grease resistance and resist aggressive fats. Their water stability makes them compatible with water-borne spray and other applications systems. Their high molecular weight allows C6 molecules a shorter, less expensive regulatory path, and imparts thermal stability. They resist hydrolytic degradation. C6 perfluoroalkyls produce very low residuals, which enhances their natural regulatory compliance.

The original process is easily altered to vary the structure and the properties of the C6 molecules. For example, the process can control the fluorinated Rf tail length and ratio; the amine reactant structure; and can vary the amine-reactive functionality and ratio of the molecule.

Because of their environmentally more friendly nature, C6 chemicals are unlikely to be regulated as C8 perfluoroalkyls have been.

GREEN CHEMICALS THAT CONTROL MICRO-ORGANISMS

A food-grade, organic-compliant, non-toxic, non-allergenic, non-hazardous and completely biodegradable anti-microbial product has been developed that also acts as natural food-safe preservative. This product offers a natural and safe solution for cleaning, sanitising, preserving food, personal care and healthcare products, agricultural growth stimulation, fresh produce treatment, hospital sanitation and new consumer product development. Businesses have to compete in an increasingly environmentally health-conscious marketplace. Focus is to produce products that are safe, do not expose consumers and communities to potentially harmful substances, and that provide functionality and value.

Scaled for commercial production, this product offers a non-toxic phyto-pharmaceutical of organic origin that acts mechanistically by destroying the cellular membrane of a micro-organism. Its efficacy is due to a synergistic reaction created between the non-toxic bioflavonoids, ascorbic acid, and other fruit acids that are included in the formulation. The final product is carried in a food-safe solvent, which makes the active ingredients infinitely soluble in water and alcohol. It is completely natural and is certified by SGS; other certifications are in renewal process. Completely harmless to the solid cells of plants, to animals, and to people, it is a terminal sanitizer and is not required to be rinsed off treated surfaces once applied. It is food-contact safe and non-hazardous, non-toxic, and non-corrosive. With a kill rate of 99.99 per cent on contact, bacteria (gram positive, gram negative, multi-resistant strains), viruses, fungi, and moulds have not been found to build up resistance to it through mutation because of its unique physical "killing." The retained efficacy and non-rinse application makes it effective against biofilms, which can be a major source of food contamination.

The process of production includes drying and grinding the pith and pulp of certain citrus fruits into a fine powder. This is mixed with natural ingredients, dissolved in water, and cold water extracted, whereafter it is pressure filtered.

The product has been proven to effectively kill the following bacteria strains:

- Bacillus cereus, bacillus subtilis;
- Clostridium perfringens, clostridium botulinum;
- Listeria monocytogenes, lactobacilus pentoaceticus;
- Propionibacterium acne;
- Staphylococcus aureus, streptococcus faecium;
- Peptostreptococcus species;
- Methicillin resistant staphylococcus aureus, vancomycin resistant enterococcus faecium;
- Escherichia coli, proteus vulgaris, campylobacter fetus, salmonella cholerasuis;
- Bacteroides species;
- Pseudomonas aeruginosa;
- Neisseria meningitidis.

The IP holder is already selling the product in numerous markets, to a variety of different industries, but is still at a very early stage of expansion. There is great interest in the product from the food, cosmetic, and general hygiene sectors, and distribution channels are currently set up. Two UK master patents are in place, and are currently in PCT stage. Various industry– or application-specific patents will support the master patents.

6

Modern Irrigation Systems

DRIP IRRIGATION SYSTEMS

Drip irrigation is today's need because Water—nature's gift to mankind is not unlimited and free forever. World water resources are fast diminishing. The one and only answer to this problem is Modern Drip Irrigation Systems.

After detailed study of inter-relationship among soil, water, crop, land terrain and related agro-climatic conditions, Jains design a suitable and economically viable system ,to deliver a measured quantity of water at the root zone of each plant at regular intervals. This is to ensure that the plants do not suffer from stress or strain of less and over watering. The system installed at the farmer's field is commissioned and training imparted to the farmer, followed by regular after sales services.

The result—a totally customised, efficient and long-life system which ensures saving in water, early maturity and a bountiful harvest, season after season, year after year. Apart from all this, savings in labour and fertiliser costs.

Why Drip Irrigation only?

- Based on careful study of all the relevant factors like land topography, soil, water, crop and agro-climatic conditions, we select the most suitable and scientific micro irrigation system.

- We at Modern Irrigation do not merely sell the micro irrigation system, we provide Agronomic and Extension support, after sales services and all technical supports for getting better crop returns. And for this, we have more than 300 technocrats, engineers, agronomists, horticulturists and regional offices, as well as trained dealers, distributors all over India and abroad.
- Modern Micro Irrigation System is made from high quality virgin raw materials, using advanced machinery. It is durable, reliable and meets International quality standards.
- Apple, grapes, banana, sugarcane, tea, coffee, cotton, mango, teak-wood, vegetables, flowers... whatever may be your crop, we have a suitable micro irrigation system for each of them. All the system components are manufactured by us in our plant at Jalgaon, under strict quality control norms at every stage of production.

Benefits of Modern Drip Irrigation Systems

- Has recorded increase in yield up to 230 per cent;
- Saves water up to 70 per cent compare to flood irrigation. More land can be irrigated with the water thus saved;
- Crop grows consistently, healthier and matures fast;
- Early maturity results in higher and faster returns on investment;
- Fertiliser use efficiency increases by 30 per cent;
- Cost of fertilisers, inter-culturing and labour use gets reduced;
- Fertiliser and Chemical Treatment can be given through Micro Irrigation System itself;
- Undulating terrains, saline, water logged, sandy & hilly lands can also be brought under productive cultivation.

Drip irrigation system delivers water to the crop using a network of mainlines, sub-mains and lateral lines with emission points spaced along their lengths. Each dripper/emitter, orifice supplies a measured, precisely controlled uniform application of water, nutrients, and other required growth substances directly into the root zone of the plant.

Water and nutrients enter the soil from the emitters, moving into the root zone of the plants through the combined forces of gravity and capillary. In this way, the plant's withdrawal of moisture and nutrients are replenished almost immediately, ensuring that the plant never suffers from water stress, thus enhancing quality, its ability to achieve optimum growth and high yield.

SPRINKLER IRRIGATION SYSTEMS

Sprinkler Irrigation is a method of applying irrigation water which is similar to rainfall. Water is distributed through a system of pipes usually by pumping. It is then sprayed into the air and irrigated entire soil surface through spray heads so that it breaks up into small water drops which fall to the ground.

Sprinklers provide efficient coverage for small to large areas and are suitable for use on all types of properties. It is also adaptable to nearly all irrigable soils since sprinklers are available in a wide range of discharge capacity.

Modern Sprinkler system is an unique irrigation system. It is designed to ensure maximum water saving, combining high quality, affordability and ease of installation. All the products are made out of high strength and chemical resistance engineering plastics to achieve functional satisfaction and to maintain cost economics.

All sprinklers undergo extensive quality testing in our well equipped state of the art lab. Moreover, performance of the products are also tested in the field to ensure uniform water distribution and higher efficiency.

Application

Suitable for almost all field crops like Wheat, Gram, Pulses as well as Vegetables, Cotton, Soya bean, Tea, Coffee, and other fodder crops.

Suitable for Residential, Industrial, Hotel, Resorts, Public and Government Enterprises, Golf Links, Race courses.

Our Sprinkler Product Range includes:

Lawn Sprinklers and Garden Pop-ups

- *Spray Pop-ups*

Features and Specifications:

- High impact and corrosion resistant engineering plastic construction;
- Heavy duty SS retraction spring;
- Anti Drain Valve to avoid puddling;
- Ratchet mechanism to adjust the spray direction;
- Separate filter at nozzle inlet;
- Wiper seal with self flushing mechanism;
- SS Screw for fine tuning of radius;
- Uniform distribution profile;
- Operating pressure 2 kg/cm^2 (28.44 psi);
- Available in 2" & 4" pop up height;
- ½" BSP female threaded bottom inlet;
- Available in full, half, quarter and three quarter circle spray patterns.

- *Nozzle for Spray Pop-ups*

Features and Specifications:

- Ultrasonically welded;
- S.S. Screw for fine tuning of radius;
- PE inlet filter to prevent clogging.

Three options for installation using spray nozzle adaptor with:

1. Multipurpose stake with adaptor for J-Bubbler

2. Stake for Bubbler
3. ½" male threaded riser.

Polyethylene inlet filter to prevent clogging.

Applications:

- Suitable for above-ground installations for the irrigation of shrubs, small lawns and gardens.

❑ *Swing Joint*

Features and Specifications:

- Stable engineering plastic construction.
- Makes the pop-up installation fast and easy.
- Flexible and leak tight swivel joints help to install pop-up in problem areas and cushions it from external impact.
- Available in ½" and ¾" BSP male threaded connection and 6", 12", 18" and 24" lengths.
- Maximum operating pressure 10 kg/cm^2 (142psi).

Applications:

- Suitable for installation of pop-ups away from pipe lines, in difficult areas like the corners or near the side wall.
- Protects the pop-up from external impact or movement due to lawn mower etc.

❑ *Flexi Pipe*

Features and Specifications:

- Manufactured from virgin Polyethylene.
- Flexible hence very easy to install.
- Available in 20 and 25 mm sizes.
- Can be used with compression adaptor and service saddle to install pop-ups.
- Maximum operating pressure 10 kg/cm^2 (142 psi).
- Available in standard coil length of 25 m. Other coil lengths can be supplied on demand.

- Flexi pipe assembly including flexi pipe (specified length) and compression fittings can be supplied on request.

Applications:

- Suitable for installation of Pop-ups.

❑ *Floppy Sprinkler*

Features and Specifications:

- Revolutionary new concept sprinkler having no rotating or moving parts.
- Unique full circle spraying pattern prevents dry patches.
- No moving parts, so maintenance free.
- Pop-up feature pops silicon tube under pressure to irrigate and pops-down at shut off.
- Light weight enables to install floppy sprinkler on heights using aluminum risers.
- Built-in flow regulator provides uniform discharge against pressure fluctuations in hilly/undulating terrain and allows longer lateral length in flat and moderate slopes.
- Large droplet size minimizes wind drift and evaporation losses giving high application efficiency.
- Operating pressure range 2-6 kg/cm^2 (28-85 psi).

Applications:

- Ideally suitable for irrigation of closely spaced crops like sugarcane, vegetables, oil seeds, onions, potato, fodder, cereals, etc.
- Suitable for landscape and turf irrigation.
- Recommended for irrigation in irregular/undulating terrain and hilly areas.

❑ *Quick Coupler*

Features and Specifications:

- Specially designed for floppy sprinkler.

- Excellent weather and chemical resistance.
- Available in:

  ½″ BSP threaded/20 mm barbed and

  ¾″ BSP threaded/25 mm barbed connections
- Can be used with Poly hose, HDPE, PVC or GI riser.
- Built in non-return valve for perfect sealing.
- Easy to handle, maintain and install in portable irrigation system.

Applications:

- Used for quick removal and replacement of floppy sprinkler during the irrigation cycle

❑ *Jain Floppy Saddle*

Features and Specifications:

- Excellent weather and chemical resistance.
- Available in barbed 20mm and 25 mm in single saddle and threaded ¾″ BSP female connection.
- Can be used with HDPE or Poly hose pipes using cable ties.

Applications:

- Threaded saddle used for fixing threaded riser pipe on the lateral and barbed saddle used for connecting poly hose to floppy sprinkler.

❑ *Quick Coupler Riser Assembly*

Features and Specifications:

- Factory assembled unit easy to handle, maintain and install, comprising.
  1. ¾″ Aluminium riser of required length.
  2. Female threaded ¾″ BSP socket to mount the floppy sprinkler.
  3. Quick coupling adaptor and saddle.

4. Two cable ties to tie up quick coupling saddle with HDPE and/or poly hose pipe.
5. Poly sleeve 600 mm long to protect from saddle plugging.

Applications:

- Used for quick removal and reinstallation of floppy sprinkler riser assembly

OVERHEAD SPRINKLERS (HIGH-LEVEL SPRINKLERS)

❑ *Impact Sprinkler-Full Circle*

Features and Specifications:

- Full circle impact sprinkler.
- Corrosion proof I-beam construction for added strength.
- Stainless steel spindle, fulcrum pin and spring.
- Well balanced gun-metal swing arm ensures smooth and uniform rotation.
- Brass nozzles with hydro-dynamically designed inner profile for precise jet and longer radius of throw.
- Easy to install and maintain.
- Uniform precipitation rate.

Applications:

- Suitable for field crops like pulses, oil seeds, vegetables, sugarcane, cotton, cereals, tea, coffee and fodder crops.

❑ *Impact Sprinkler - Part Circle*

Features and Specifications:

- Adjustable arc impact sprinkler.
- Corrosion proof I-beam construction for added strength.
- Stainless steel spindle, fulcrum pin and spring.
- Easy arc adjustment from 30º to 360º.
- Strong, well balanced gunmetal swing arm ensures smooth and uniform rotation.

- Brass nozzles with hydro-dynamic inner profile for precise jet and longer radius of throw.
- Easy to install and maintain.
- Uniform precipitation rate.

Applications:

- Can be efficiently used where field layout is uneven or at the sides and corners of the field.
- Suitable for field crops like pulses, oil seeds, vegetables, sugarcane, cotton, cereals, tea, coffee and fodder crops.

❑ *Komet Medium Volume Rain Gun Model - 162 and 163*

Features and Specifications:

- Heavy duty construction.
- Most suitable for a variety of climates like tropical, temperate and humid climates as in India.
- Light in weight and easy to install.
- Long wear life and low maintenance.
- Uniform distribution profile with adjustable jet breaker arrangement.
- Better pattern coverage and good performance in windy conditions.
- Suitable for both solid set field irrigation and mechanically moved (centre pivot etc.) systems.
- Various interchangeable nozzles provide better control in watering and irrigation of desired area and crops.
- Reduced risk of structural damage of soil and delicate plants due to evenness and lightness of watering.
- Available in full circle and part circle models.

Applications:

- Recommended for field crops like sugarcane, pulses, oil seeds, cereals, tea, coffee, and vegetables.

- *Komet Big Volume Rain Gun Model Twin 95 Plus*

  Features and Specifications:

- Heavy duty construction using high-tech composite materials, in conjunction with sturdy drive arm.
- Extraordinary operation at all pressure levels, works smoothly and uniformly over whole nozzle pressure range.
- Excellent hydraulic design, large barrel cross section and full size taper bore nozzle allows for maximum possible throw and performance.
- Maintenance free, self lubricating sealed bearing, instead of ball bearings allows for long reliable operation over time.
- Three stream break-up frequencies can be selected manually, without tools, in order to adjust rotation speed and to optimise the water distribution.
- Unique patented self compensating break system keeps overall performance constant over time.
- Intermittent dynamic jet breaker option, increases the stream diffusion which allows a more delicate irrigation for sensitive crop and provides flexibility in operation at lower pressure.
- Patented drive system with excellent stream diffusion allows for smooth and steady operation.
- At start up of the system the gun produces a fan-type spray reducing to a minimum crop damage and soil erosion due to run off.
- Light weight, puts less stress on irrigation system structures.
- Available in full circle and part/full circle models with seven interchangeable nozzles to meet discharge and radius requirements.

  Applications:

- Recommended for field crops like sugarcane, pulses, oil seeds, cereals, tea, coffee and vegetables etc.

- Easy to use with portable irrigation system.
- Useful for large turfs, lawns and playground.

❑ *Komet Big Volume Rain Gun Model Twin 140 Plus*

Features and Specifications:

- Heavy duty construction using high-tech composite materials, in conjunction with sturdy drive arm.
- Extraordinary operation at all pressure levels, works smoothly and uniformly over whole nozzle pressure range.
- Excellent hydraulic design, large barrel cross section and full size taper bore nozzle allows maximum possible throw and performance.
- Maintenance free, self lubricating sealed bearing, instead of ball bearings allows for long reliable operation over time.
- Three stream break-up frequencies can be selected manually, without tools, in order to adjust rotation speed and to optimise the water distribution.
- Unique patented self compensating break system keeps overall performance constant over time.
- Intermittent dynamic jet breaker option, increases the stream diffusion which allows a more delicate irrigation for sensitive crops and provides flexibility in operation at lower pressure.
- Patented drive system with excellent stream diffusion allows for smooth and steady operation.
- At start up of the system the gun produces a fan-type spray reducing to a minimum crop damage and soil erosion due to run off.
- Variable trajectory angle option is also available which allows you to change trajectory angle from 100 to 280 thus provides more uniformity even in windy conditions.
- Special high trajectory model with 440 trajectory is available for dust suppression.

- Eight interchangeable nozzles to meet the discharge and radius requirements.
- Light weight, puts less stress on irrigation system structures.
- Available in full/part circle models with seven interchangeable nozzles to meet discharge and radius requirements.

Applications:

- Recommended for field crops like Sugarcane, Pulses, Oil Seeds, Cereals, Tea, Coffee & Vegetables etc.
- Easy to use with portable irrigation system.
- Useful for large turfs, lawns and playground.

7

The Impacts of Coca Cultivation and Coca Eradication

INTRODUCTION

Lowland tropical forests along the eastern margins of the Andes (the Tropical Andes biodiversity hotspot: Myers 1988, 2003, Myers et al. 2000) have witnessed waves of migrants during the late 20th Century who have converted forest to cropland and pasture. Aided by a plethora of policy initiatives, they are fragmenting lowland humid and lower montane tropical forests; threatening habitat integrity; compromising ecosystem services; and impacting their own livelihoods and those of indigenous people. These phenomena have been mapped in Colombia (Viña et al. 2004, Etter et al. 2006), Ecuador (Walsh et al. 2003, Pan et al. 2004), Peru (Naughton–Treves 2004), and Bolivia (Steininger et al. 2002, Millington et al. 2003, van Gils and Loza Armand Ugon 2007).

Since the late 1970s, an important income source for many colonists has been coca-Erythroxylum novogranatense (D. Morris) Hieron. var. novogranatense in Colombia (MacGregor 1993, Fjeldså et al. 2005) and Erythroxylum coca Lam. var. coca in Bolivia (South 1977, Sanabria 2004) and Peru (Young 2004). Traditionally coca leaves have been chewed in the Andes (Commission of Enquiry on the Coca Leaf 1949,

Meruvia 2000), but they are also the base ingredient for cocaine paste (MacGregor 1993). Since the 1970s, rising demand for cocaine in the developed world (Walsh 2004) has maintained high prices for coca relative to other crops, making it a lucrative cash crop. Perceived negative social impacts of coca chewing stimulated the UN Commission of Enquiry on the Coca Leaf in the 1940s (Commission of Enquiry on the Coca Leaf 1949). High levels of cocaine production and trafficking in South America by the late 1950s prompted the Interamerican Conference on the Illicit Traffic in Cocaine and Coca Leaves in 1960 and subsequent legislation (Parreiras 1961). Continued production and trafficking (Joseph Rowntree Foundation 2000, Steinberg et al. 2004) forced the U.S. government to initiate the "War on Drugs" and the enforcement of anti-coca policies during the 1980s and 1990s (McCoy 2004), which included uprooting and burning bushes, spraying defoliants, and introducing alternative development programmes that promote substitute crops. Amazon deforestation caused by ranching and crop cultivation have been described and modeled elsewhere (e.g., Moran et al. 1994, McCracken et al. 1999, Mäki et al. 2002, Walker 2004); however, in the presence of illegal coca, agents and drivers of deforestation are also influenced by coca cultivation, cocaine processing, and policies designed to counteract coca-related activities (Bradley and Millington 2008). Social and political issues related to coca have generated much research, but little attention has been paid to environmental impacts. These include deforestation, forest fragmentation, encroachment onto conservation units, soil erosion, and point pollution of soil and water (Dourojeanni 1992, MacGregor 1993, Young 1996). The enforcement of anti-coca policies further degrades environments in coca-source regions, e.g., defoliants contaminate soil (MacGregor 1993), streams, and groundwater.

Kaimowitz (1997) theorised that deforestation rates would be low under coca-dominated production systems because of the high labour requirements needed to maintain more than a hectare of coca bushes. Henkel (1995) presented a theoretical, economic argument based on calculations of cultivated areas in Chapare. He argued that without coca cultivation only 20 per cent of farmers could be supported on the 114 000 ha of land cleared by 1990, and to support the remaining four-fifths, 700000 ha of forest would have to be cleared. This equated to an average expansion of each farm from a few to 20 ha. To our knowledge, no detailed, quantitative analysis of deforestation rates under coca production and anti-coca policy regimes has been published.

THE CHAPARE COLONISATION ZONE

The Chapare colonization zone covers approximately 24 500 km^2 of central Bolivia and, after parts of Colombia and Peru, it is the third largest coca-source region globally United Nations Office on Drugs and Crime (UNODC 2006). Elevation ranges from 200–500 m.a.s.l. and, before colonisation, it comprised terra firma and inundated lowland forests on alluvial fans and seasonally inundated floodplains (Montes de Oca 1989). The region is connected by road to the second (Santa Cruz) and third (Cochabamba) largest urban areas in Bolivia: both are major markets for Chapare's produce. The Andes, to the south, are covered with species-rich, tropical montane forests containing high levels of endemism. Chapare forms a transition zone between biologically important Andean montane forests and the Amazonian humid tropical lowland forests to the north (Myers 1988, Myers et al. 2000, Navarro et al. 2004), and biological interchange between the two is compromised by development in Chapare.

Colonisation began in 1910 at Todos Santos in central Chapare. The first government-sponsored schemes were initiated in the same area a decade later (Lavandez 1925, Weil 1983). Until the 1960s, colonisation was generally unsuccessful (Mather 1922, Crist 1946, Weeks 1946, Edelmann 1967, Fifer 1967, Weil 1983). However, during the 1960s, several hundred communities and some large-scale commercial ventures were established in central Chapare under Plan Decinal. They caused forest destruction that has been attributed to farmers inexperienced in humid tropical environments and weak government support (Rivera 1990). Driven by national shortages of potatoes and maize, colonists invested in rice, leading to rapid deforestation (Weil and Weil 1983). By the 1970s, fruit cultivation and ranching were commonplace. At this time, coca, which grows well in Chapare and is easy to transport as dried leaves, had become popular, although Rodriguez (1965) noted it was cultivated in the late 1950s and early 1960s. During the 1980s, colonisation shifted to eastern Chapare along a new road linking Cochabamba and Santa Cruz (Millington et al. 2003). Improvements in transport infrastructure should have favoured the growth of marketable, perishable crops, but coca continued to dominate because of the high prices created by strong global demand for cocaine (Bostwick et al. 1990). Coca production became so prevalent that the United States forced the Bolivian government to implement anti-narcotics policies (Sanabria 2004). To comply with U.S. policies, the Bolivian government set coca-field destruction targets that were accompanied by increasingly aggressive eradication campaigns conducted by the Bolivian police and military during the 1980s and 1990s (Morales 2004). Law 1008 (Ley del Regimen de la Coca y Sustancias Controladas) in 1988 recognised legal cultivation in traditional coca-growing areas in subtropical valleys of the Andes and made cultivation in the Chapare lowlands illegal. Subsequent anti-coca packages included alternative development programmes built around crop substitution programmes supported by U.S. Agency for International Development (USAID) marketing campaigns and subsidies. Many substitute crops failed to reward farmers adequately because of: (i) lower economic returns than those from coca; (ii) poorly researched markets; (iii) markets that became oversaturated and crashed (Hellin 2001); and (iv) lower economic returns for substitute crops than anticipated because of middlemen-dominated transport networks and wholesale markets. These factors undermined confidence in substitute programmes generally (Jackson et al. 2003), and in spice and palm heart cultivation specifically (Hellin 2001). Returning to coca was a common response.

CHRONOLOGICAL ORDER OF SETTLEMENT

We used similar methods to Rindfuss et al. (2003). They included a time series of satellite images, household and farm surveys, and analyses of archived cadastral surveys, agricultural statistics, and economic data. Our time frame was 1963 to 2003. We selected three distinctly different communities in central and eastern Chapare that were colonized at different times for detailed analysis. Each represents one of the main types of agricultural activity in the zone. Their names have been changed to Communities I, II, and III, which also refers to the chronological order of settlement.

Land use and forest conversion were mapped from a time series of Corona, Landsat MSS, TM and ETM+ satellite images. Corona data are panchromatic high spatial resolution declassified U.S. spy satellite photographs acquired between 1959 and 1980, which have begun to be used to extend land-change studies backward into the 1960s (Hamandawana et al. 2007, Tappan et al. 2000). We used Corona KH4A data acquired on 23 September 1966 (ground resolution ?3 m) for Community I. Cloud-free Landsat data for Communities I and III (Path/Row 232/72) were acquired on 1 July 1976, 4 October 1986, 18 July 1993, 18 February 1996, and 14 July 2000; and for Community II (Path/Row 231/72) on 16 September 1975, 25 July 1986, 9 July 1992, 5 August 1996, and 5 June 2000. We geo-rectified the images (RMS <0.3 pixels) and mapped 10 land-cover classes using the ISODATA clustering algorithm in Erdas Imagine (Bradley 2005). We masked water bodies and shadow, and excluded them from the analysis. We used information obtained during field verification campaigns and interviews with farmers to allocate each cluster class to either a forest or a non-forest class for each image in the series. When we could not resolve a cluster class on the basis of this information, or contextually, it was allocated to forest. A forest/non-forest map was created for each year in the image sequences. Because we allocated ambiguous land covers to the forest class, forest clearance metrics are conservative and deforestation rates may be slightly underestimated. We digitised cadastral plans for each community and overlaid them on the forest/non-forest maps. We calculated forest and non-forest areas from each image map, and the proportions of forest remaining in ArcGIS. Rates of forest loss were calculated for inter-image periods using the following equation:

$rj = (CAk1j - CAk2j) / (k1 - k2)$

where r = rate of clearance, CA = cleared area, j = community and k = time point. Rates of forest clearance are assumed constant between image dates.

We conducted household surveys in each community in 2000 and 2003. These focussed on past and current land uses, with forest/non-forest maps being used as prompts. Cultivating coca in Chapare is clandestine, and talking to farmers about it is challenging. Therefore, we did not broach the subject until we had gained their confidence. It is too risky to enter farms without an invitation or an escort, so only if a farmer agreed did we walk their farms with them. Even then, we did not make detailed maps of coca cultivation using GPS during field visits as this would have jeopardised our research. Rather, they drew sketch maps of their past and current land uses, recording locations of specific crops and agroforestry complexes. We compared the sketch maps with the forest/non-forest maps to verify past forest cover and land use, i.e., to convert them to land-use maps. Some cropped areas were clear on the imagery, but others were less so because they were intercropped or overgrown.

We interviewed 14 of 110, 16 of 90, and 13 of 60 households in Communities I, II, and III, respectively, to provide information pertinent to Kaimowitz's (1997) labour argument. Using interview data and land-use maps, we created chronologies of dominant farming activities for each community. We divided the 43 farms researched into six forest clearance typologies, which were based on analyses of the trajectories of farming activities obtained from interviews and farm walks, and then identified the rates of clearance associated with each typology to define them in terms of land-use trajectories and rates of forest loss.

We obtained economic data from USAID. These are discontinuous as a result of short-term programme funding, and are only available for the last two decades. Data from the Chapare Regional Development Project (CORDEP; 1993–1996) is for wholesale prices of substitute crops in the major cities.

Price data on alternative crops (2001–2003) from the Counter-Narcotics Consolidation of Alternative Development (CONCADE) programme were collected at truck checkpoints on the borders of Chapare. We also obtained data on farmgate prices during interviews, and time-series of coca prices and areas of under coca cultivation from the Bolivian National Statistics Institute (INE) and the UNODC. These data were used to contextualise trends in dominant farming activities and forest clearance rates, and test Henkel's (1995) ideas on comparative prices and clearance.

THE RATES OF FOREST CLEARANCE

Complete forest cover in Community I in 1963 had declined to 33 per cent by 1986, although it had recovered to 44.8 per cent by 2000. In Community II, forest cover declined from complete cover before colonisation (1972) to 42.8 per cent by 2000. Over the same time, forest cover in Community III fell from complete cover to 53 per cent.

Most farmers in Community II settled in 1983 to form a cattle cooperative, nonetheless between 1981 and 1995, a high proportion (34%-62%) grew coca. That proportion fell, and by 2000, when only 12 per cent admitted cultivating it, most (94%) were clearing forest for pasture. Rates of clearance were highest immediately after the two periods of settlement, in 1972 (due to rice cultivation) and 1983 (due to subsistence crops, coca, and pasture). The relationships between the rates of forest clearance and the proportion of farmers cultivating coca or ranching are not clear; however, the greatest increase in the mean cultivated area occurred between 1987 and 1993 when an additional 44 per cent of farmers began to clear forest for pasture.

It is clear that high rates of forest clearance were prevalent in Community III between 1983 and 1986 when farmers who had recently moved there cleared forest to grow subsistence crops and coca. After 1990, the proportion of farmers engaged in coca decreased. This was mirrored by an increase in the proportion of farmers cultivating substitute crops, which in this community were dominated by bananas. This switch occurred because farmers were forced to find alternatives to coca, and had joined USAID substitute programmes. The relationships between forest clearance rates and periods of coca cultivation and substitute crops are clearer in Community III than in the other communities.

In general, where coca was cultivated, forest clearance rates tended to be low, but increased as coca substitutes became dominant. Clearance rates were normally high when cattle ranching was the major activity. Some clearance typologies of the six identified were more frequent that others, e.g., in Community III, 76.9 per cent of farms had trajectory 4, coca to substitutes with slow to fast clearance, and in Community II, 62.5 per cent of ranches were predominantly for cattle, with a period of coca when forest was cleared more slowly.

THE ANALYSIS OF FOREST

The analysis of forest/non-forest maps combined with interviews enabled contemporary and past land uses to be established. We contextualised these by reference to the time-series of crop price data to develop an understanding of the economic rationales behind farmers' decisions to clear forest at different rates under coca and anti-coca regimes. Although coca prices were at their lowest between the mid 1980s and the late 1990s, the farmers we interviewed strongly argued that risks associated with illegal coca cultivation during the period of low prices were worth taking because coca provided a better selling return than other crops. In part this is because, regardless of price, coca can provide three harvests a year in Chapare unlike most other crops.

To help develop this economic argument further, wholesale prices for pineapple, a major substitute crop, between 1992 and 2003. The gaps in these data emphasise the difficulties in acquiring data on crop production and prices in Bolivia. Unlike coca, pineapple only provides one crop a year. The investments in pineapple are worse still compared with coca because the first harvest is delayed until 2 years after planting, after which fields are converted to other land uses. Pineapple prices also have a well-developed seasonal trend unlike coca, which maintains a more-or-less consistent price throughout the year.

Peak prices for pineapple declined significantly between the mid 1990s and 2000-2003. It is thought that this was the result of market controls through subsidies. Subsidies also led to stabilisation of banana prices between 2000 and 2003 compared with the mid 1990s when, although prices increased, their behavior was spiky. Compared with pineapples and bananas, a major advantage that stable coca prices gave farmers, even when coca prices were low, was a predictable, assured household income. We argue this led to the development of risk-adverse behaviour amongst farmers, which made them reluctant to take risks with substitute crops later unless they could be assured of a secure market or a crop subsidy. The strongly advantageous selling returns of substitute crops reinforced this risk-adverse mentality.

These economic arguments are highly pertinent to our consideration of deforestation rates under coca and anti-coca regimes. In the theoretical example, it is clear that the high economic returns of coca would not only strongly influence a farmer's decision to grow coca, but would impact on the amounts of forest that would need to be cleared to grow coca or a substitute crop. In the example, 14 ha of palm need to be grown to produce palm hearts equal in monetary value to a hectare of coca. These data strongly indicate that farmers would generally choose to cultivate coca over substitute crops because of its higher selling return and its relative stability compared with the unstable prices and markets for most substitute crops.

However, this has not been the case all the time in Chapare. The economic arguments above generally only apply when Chapare was under a coca regime, i.e., periods when coca prices were high and/or coca eradication was weakly enforced. It is at these times that deforestation rates have been low: mean forest losses on the farms we have studied have varied from negative values indicating forest regrowth to losses of around 0.4 ha/yr. The economic arguments do not apply to the periods we term anti-coca regimes, times when the implementation of anti-narcotics policies have overcome the economic imperatives of coca growing and forest conversion rates increased rapidly (mean forest clearance rates on farms range from 0.9–1.1 ha/yr). Therefore, although Kaimowitz's (1997) and Henkel's (1995) economic arguments have some value in explaining deforestation rates in Chapare, they underplay the importance of policy shifts.

The impacts of changing policies on the areas of coca cultivated and eradicated, and substitute crops grown between 1980 and 2000, Two communities—I and II—were settled before 1980, the approximate date when the global cocaine market became serviced by coca leaves from Chapare and the area cultivated increased markedly. Our evidence indicates that either rice or fruit were cultivated by farmers before 1980 and that relatively little coca was grown. The policy that drove land use in the 1960s was the Plan Decinal, but weak markets and a poor infrastructure led farmers to begin cultivating coca in large quantities from the late 1970s onward because of its comparative economic advantage and lack of government controls on its cultivation. During the 1980s, the existing farmed areas in Communities I and II, and the newly cleared areas in Community III, were dominated by coca. Eventually policies to counter the illicit coca trade were triggered. Despite moves to discourage cultivation between the mid 1980s and 1991, data clearly show that for the remainder of the 1980s coca was dominant and the area under substitute crops only increased slowly. This trend is mirrored in the deforestation dynamics of the communities we researched. If the high rates of clearance for subsistence cultivation are ignored, the mean rates of forest conversion to agriculture across communities were low when coca was the main economic activity, e.g., Community I (-16.88 ha/yr) and Community III (8.2 ha/yr). Community II maintained high rates because it is a predominantly cattle-rearing community, compared with field and orchard crop cultivation in the other two communities, and clearance for pasture generates high rates of forest conversion.

Implementation of anti-coca policies strengthened significantly after 1986, and eventually the area under substitute crop programmes increased, e.g., from 60 000 to 120 000 ha between 1993 and 2000. These broad land-use trends apply to all three communities. In Community I, fruit cultivation again became important, and in Community III, farmers began to cultivate bananas, black pepper, palm hearts, passion fruit, and pineapples. The responses are clearly evident in the mean community forest clearance rates, which increased to 30.98 ha/yr. in Community I between 1993 and 1996, and 57.5 ha/yr in Community III after 1996. The situation in Community II—the cattle-rearing community—was different in that farmers had continued to clear large areas of forest for pasture, and during the implementation of anti-coca policies, overall community clearance rates remained high at over 100 ha/yr.

By combining the economic and policy data with the analyses of deforestation dynamics at community and farm levels between the 1960s and 2003, we have identifed three epochs of colonist agriculture, each with a distinct deforestation dynamic. Epoch I was characterized by the cultivation of conventional tropical crops and is mainly restricted to the decades before 1980. Deforestation rates were high, particularly in the initial few years when farmers were establishing their farms. In Epoch II, which extends throughout the 1980s and into the mid 1990s, land-conservative coca dominated agricultural practices and forest clearance rates plummeted. We call this the coca regime. From the mid 1990s to the end of our survey, Epoch III was characterised by the cultivation of alternatives to coca. During this time, land-hungry substitute crops necessitated the clearance of much forest and forest losses increased. This we term the anti-coca regime.

Having, for the first time, quantified rates of forest conversion under coca and anti-coca regimes, it is possible to revisit the arguments postulated by Kaimowitz (1997) and Henkel (1995) about deforestation under both regimes. In terms of the household labour constraint argument, we have evidence from the sequences of forest/non-forest maps for individual farms contextualised with interview data. In the few areas where farmers continued to grow coca in the face of anti-coca policies (Trajectory 1, 5), labour was directed toward coca and little was available to clear forest. Forest clearance rates on such farms were very low. More common is the property, which is owned by a farmer who volunteered to give up coca and grow substitutes. Growing these proved to have lower labour requirements than tending coca bushes, and forest was cleared faster. Although the two examples lend some quantitative support to Kaimowitz's (1997) argument, it is difficult to evade the economic arguments that probably also prevailed in these two households. His argument is further weakened by evidence from Community II, the community where coca was always secondary to cattle. Here, long-term decisions were made to channel household labor into creating pasture so that herd sizes could be increased. When coca was cultivated in Epoch II, it was very much an income supplement. Typical is trajectory 3, which is restricted to farmers in Communities I and II who had cattle. High rates of forest conversion were maintained throughout Epoch II on these farms/ranches because of the long-term goal of building up herds. In Community I, the farm labor pool had been fully occupied cultivating conventional crops for almost two decades during Epoch I, during which time forest clearance rates were high. Clearance rates decreased in the mid to late 1980s (Epoch II) because, although farmers grew conventional crops as well as coca, most household labor was diverted to coca cultivation. Labour resources were too stretched to clear more forest, but additionally our interviews indicate that simultaneously other factors promoted the decline in forest clearance rates. These were (i) that there was little or no forest left to clear in some plots; (ii) low yields had led farmers to abandon some crops leading to forest regrowth; and (iii) some citrus stands were maturing and appearing as forest regrowth on image-derived forest/non-forest maps. In places where farming activities were well-established and/or farmers had developed long-term strategies, e.g., in Communities I and II, we cannot readily attribute changes in forest clearance rates to household labor constraints created by the coca economy. The explanations underpinning changes in deforestation rates in such locales are more complex. In Community III, however, where the initial settlement coincided with the coca regime of Epoch II, there is a stronger case that the clearance rates were influenced by labour constraints.

The interviews provide strong support for Henkel's (1995) economic argument concerning forest clearance. Most farmers unambiguously confirmed that they had cleared more forest after they had given up coca (Epoch III) to achieve acceptable economic returns on investments in substitute crops and to maintain household incomes. Small areas of coca were grown during Epoch II, indicating that households dependent on coca were conservative deforesters. Enforcement of eradication and promotion of substitutes led to more forest being cleared, particularly among large families. In some circumstances, the image interpretation confirmed what the farmers suggested in interviews. For example, in Community III, where there was widespread replacement of coca with substitute crops, average farm sizes have increased by about 7 ha (in plots no larger than 20 ha) in Epoch III. This is much less than that predicted by Henkel, but we note that many farmers were cultivating coca alongside substitute crops until recently, and some possibly continue to do so. For those who chose substitutes over coca, the consumption of forest is all too clear from image analysis and walking through farms. As already indicated, coca cultivation in Community II had little impact on the rate of forest-to-pasture conversion between Epochs I and II and farms continued to expand by approximately 1 ha/yr. In terms of Henkel's (1995) argument, the supplementary nature of coca in Community II is pertinent, but its location is also germane to the argument. Located in western Chapare, it was the community farthest away from the coca heartland of central Chapare. It is, perhaps, better thought of as being at the periphery of the well-established agricultural hinterland of Santa Cruz. Consequently markets were better developed and more stable, particularly for beef and dairy. Although there is no documented meat price data for Chapare, farmers recalled that cattle prices were generally stable and rarely fluctuated through the 1980s and 1990s, eliminating one of the incentives to invest in coca. Even when the economic drivers encouraged farmers to cultivate coca elsewhere, long-term economic objectives seem to have prevailed in this community. In Community I, the rice and fruit cultivation system established before Epoch II never completely disappeared, even though forest clearance rates fell as more coca was cultivated. As a consequence, during Epoch III, farmers often reverted to their pre-coca crops and rates of clearance increased as either primary forest or forest regrowth was cleared.

8

Phytomonitoring in Agro-Technology

INTRODUCTION TO PHYTOMONITORING

The direct monitoring of growing plants, aimed at improvement of the controllable crop factors, is becoming a world standard in precision agriculture, which will change the way Growers will operate in coming decades.

The Phytomonitoring methodology incorporates plant-sensing techniques, sampling rules, measurement protocols, data interpretation and crop-specific application techniques. The total concept is embedded in modern electronics, data acquisition and communication systems, software and Internet enabled interface.

The system is tailored to the specific world of Growers, in the sense that it is: simple to operate and maintain, can be installed using simple building blocks, can be configured for each Grower demands and is moderately priced.

Phytomonitoring systems are field proven. They have been met with immense enthusiasm by agronomists and farmers in Israel and have been adapted and currently are supported by the Ministry of Agriculture in the country. Records of increased productivity and operational savings have clearly be recorded and translated into hard monetary terms.

A grower applies treatments and provides irrigation, fertigation and other manageable environmental regimes. Usually, it is based on direct observations (made also with the use of instruments) and available laboratory tests. The Phytomonitor has become an extraordinarily responsive and informative real-time feedback channel between a grower and cultivated crops, supplementing and redoubling traditional methods.

The main sections of this document covers the issues of:

PLACE AND ROLE OF THE PHYTOMONITORING IN MODERN HORTICULTURE

Modern agriculture requires customised soil and crop management to enhance the specific conditions within an agricultural field. To meet this demand, a new term, PRECISION AGRICULTURE, is used to define technologies that fully support tailored agricultural practices aimed at higher efficiency and lower impact to the environment. By right, Phytomonitoring technology belongs to the PRECISION AGRICULTURE category.

Phytomonitoring is a specialised real-time information system for horticulture. The purpose of its application is to derive new crop-related information for supporting decision-making processes for climate (where applicable) and irrigation control (Fig. 8.1).

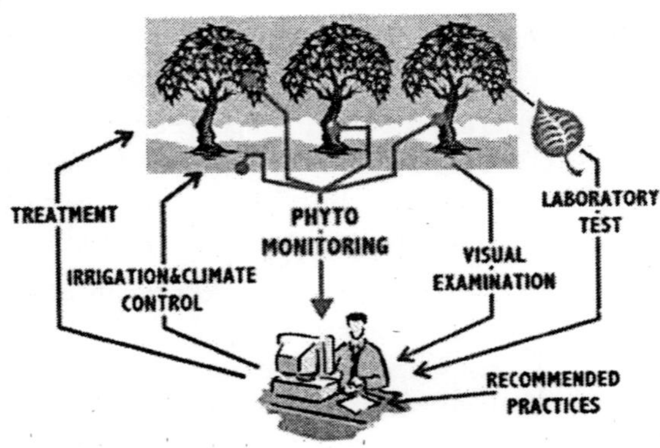

Fig. 8.1: SEQ Figure* ARABIC

By definition, phytomonitoring is a management information system for crop growing. There are three main functions of the phytomonitoring information system:

1. Standard reporting

The system can generate a customised set of measured values and their derivatives, used by a grower in daily control practice.

2. Exception reporting (warning)

The system enables the early detection of unexpected disorders in plants. This function is based on a variety of phytomonitoring indicators of plant physiological disorder.

3. Decision-support system

It enables the fine tuning of climate and irrigation regimes by trial-end-error approach. Extremely high sensitivity and short response time of the phytomonitoring channels eliminate the risk of crop damage. A grower can make a little change in a control regime and then he may get a detectable response of plants within 1-2 days. It gives him the high chances to keep only a single variable factor during a trial and also allows to prevent considerable impact in crop state. Phytomonitoring dynamic indicators of plant improvement or deterioration are necessary attributes of such decision-support system.

Due to the dynamic and relative character of the phytomonitoring observations, they have high representativeness at limited number of samples. Practically, only a few sample plants are necessary for effective monitoring of large plantations.

PHYTOMONITORING TECHNIQUE

Phytomonitoring technique is crucial for performing the following practical tasks:

- To detect plant physiological disorders at early stages of their development. Probable causes are the following: improper climate control (if available), irrigation or fertigation regimes, diseases, and pests.

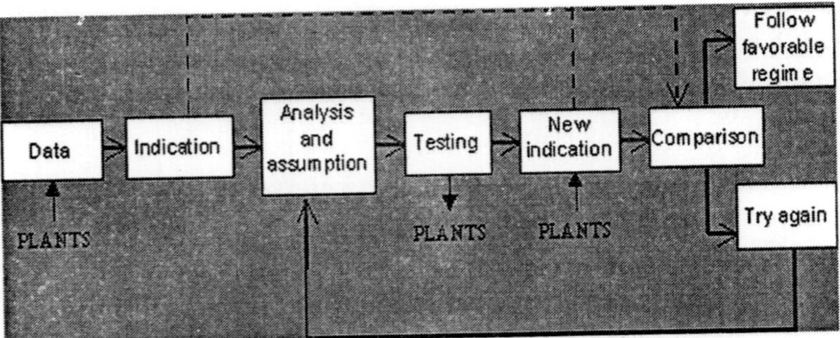

Fig. 8.2: The Phytomonitoring trial-and-error approach for tuning control regimes. The collected data are converted into plant status indications. A Grower analyses the indications and makes an assumption about subsequent modification of control regime. Further testing of the new regime brings to a new indication of plant status. In many cases, a grower is able to interpret comparison of both indications in terms of positive or negative response to a new regime. Phytomonitoring technique is now rapidly being recognised as a new standard precision farming tool for tuning cultivation regimes in greenhouses and open field crops.

- To locate the source and to identify possible causes of crop problems.
- To disclose the crop physiological response to any environmental changes in short time. It may help a Grower to examine his attempts at improving crop performance or to eliminate problematic crop factors. It also enables a Grower to tune climate, irrigation and fertigation regimes and treatments in a trial-and-error approach.
- To present a Grower both conventional and original integrated and accumulated long-term characteristics of the crop, such as Thermal time (degree/days), Daily integral of solar radiation, Reference evapotranspiration, Leaf surface wetness duration, as well as duration and magnitude of measurable stress conditions (air drought, heat, chilling, soil water stress, etc.)

This information is of great importance for estimating and planning the control policy.

Thus, using phytomonitoring technique, a grower becomes able to examine crop conditions in trial-and-error regime and, consequently, to improve crop performance and production during the very first seasons. The phytomonitoring technique may help to avoid rough technological mistakes easily. Also it allows disclosing hardly detectable, accumulated physiological disorders. And in the long run, the phytomonitor enables permanent tuning of climate and irrigation control for better crops and lower costs. It is also a perfect tool for comparative examination of different treatments and materials.

Fig. 8.3: THE UNIQUENESS OF THE PMS is provided by its complex character; combining hardware (sensors, data logging and communication facilities), software (data logging and graphics) and application techniques (selecting a rational set of sensors, allocation of sensors, data interpretation techniques and decision-support procedures).

Phytomonitoring™ is thus defined as an Information and Decision-Support System for Crop production, delivering a complete solution for plant monitoring applications.

Fig. 8.4: PhyTalkTM is an advanced system for remote wireless sensors that monitor plants (Phytomonitoring™). The system collects physiological and environmental data directly from the plant using enhanced sensors that communicate the data to software that analyses and presents it in a useful and user-friendly format. The product provides crop managers and other specialists and consultants with a quique took to collect data, analyse it, explore trends, and alert the grower when stress conditions develop.

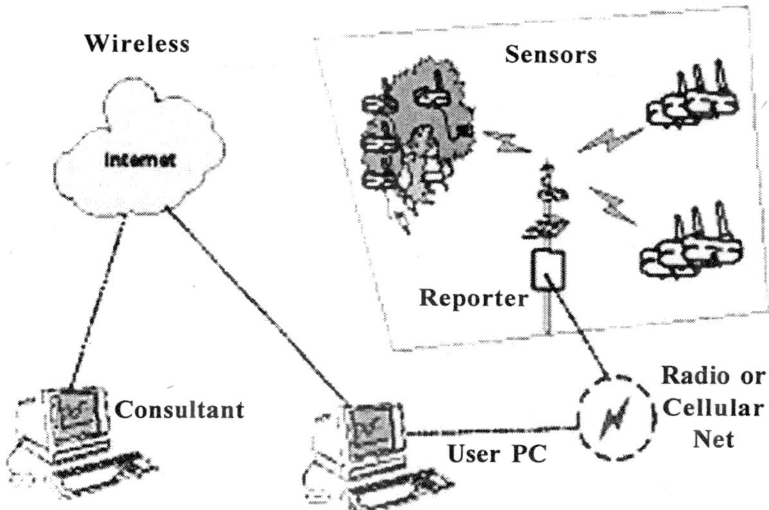

Fig. 8.5: Phytalk System Wireless Plant Monitoring System (PMS) PhyTech is a pioneer and a leader in Phytomonitoring™; the practice of Remote Sensing and Monitoring of Plants. Our advanced solutions use innovative software for collecting and analysing data, state-of-the-art sensors, and wireless communication. Our systems improve crop production and reduce cultivation costs by providing timely and accurate information regarding the physiological status of the plants and identifying stress conditions before they impact the plant or the fruit.

THE PHYTALK SYSTEM

PhyTalk is designed as a modular, distributed system. New sensors can be added as needed. The architecture supports cost effective deployment of sensors over large geographical area. The system consists of 4 components:

Sensors: Measure a variety of parameters, providing relevant and precise information about plant and environmental conditions.

Communications: Supports cordless deployment of sensors in the field and wireless communication to remote computer anywhere in the world.

Software: Developed by experts in the field of Phytomonitoring, the software support the analysis and interpretation of data in real time by displaying it in easy to read formats such as graphs and tables and supporting disorder prevention.

Application Techniques: Recommended measurement protocols and decision-support procedures for major horticultural crops.

Role of Phytomonitoring in Agriculture

1. *What is a role of phytomonitoring technology in the process of plant growing?*

Phytomonitoring is an information technology that provides a grower with invaluable information about dynamics of the crop physiological status. A picture below clearly represents a role and place of phytomonitoring in the process of plant growing. Thus, the phytomonitoring is one of the crop supervising tools, which is mostly touchy and quick for estimating a crop status trend. One may distinguish two may function of the phytomonitoring: (1) warning a grower in case of upcoming problems with the crop long before they become visible and irreversible and (2) examining crop treatments and control operations in respect of whether they improve or not the crop conditions.

2. *How many plants are used for phytomonitoring?*

The phytomonitoring interpretation technique is mainly based on the analysis of trends, which are much more representative for uniform crop than absolute crop characteristics. For example, instant temperature of several leaves may differ in a few degrees centigrade while the response to lower humidity will be uniform: elevating transpiration will reduce temperature of all leaves. Usually a phytomonitor set incorporates a single set of environmental sensors and double set of main plant related sensors. It means that only two plants may be used as driving ones at a uniform plot. By the way it gives a unique opportunity to test some new treatments or regimes only on driving plants at first, before application on a whole crop.

3. *How to learn the phytomonitoring interpretation technique?*

Every phytomonitor is supplied with application software tools, which automatically recognise many typical disorders in plants. Besides, a user may take advantage of the Reference Guide for more detailed information about phytomonitoring diagnostic technique. Any legal user may also become a member of the Phytomonitoring Club to have access to phytomonitoring data banks and having direct support of top experts and colleagues. In addition, any user may take a special training course organised by the PhyTech Ltd. and local dealers twice a year. Send a request for information about the next course at Error! Bookmark not defined.

4. *How to purchase, install and maintain the phytomonitor?*

Purchasing. Call your local authorised dealer. A list of Phytechs representatives is available. You may also call directly to the main office (Contact Us). You may also get free consultations regarding a delivery set which suits your crop. Installation. A Companys technician provides installation in most cases. For example, a routine installation session of the LPS-03MA Standard Phytomonitor includes the following steps: running a 4-wire cable between the phytomonitor location and PC. It is recommended to do it yourself beforehand; — installing the phytomonitor (about 30 minutes); — installing software (about 15 minutes); — introductory instruction (45 minutes). Maintenance. There is only one concern of a user — to keep sensors properly positioned. Usually it requires the following: to check positioning of 2-3 plant related sensors (LT-1,SF-4/5 and FI-3 if present) twice a week; to check positioning of all other sensors once in a month. All these actions take not more than an hour per month.

5. *What is expected benefit of using the phytomonitor?*

It is quite difficult to calculate correctly a benefit provided separately by an information tool, which gives a grower the management and time-critical information on his crop. Considering only a watchdog function of the phytomonitor one may suggest the following. It is known that crop yield may be only 22 per cent of what it might be because of severe stress. Assuming that the phytomonitoring watchdog can save only 5 per cent of yield, one may get approximately $17,000 per hectare for vegetable and cut-flower crops wher total gross income is estimated as $32-$37 per m²*, and up to $30,000 per hectare of pot-plants greenhouse where total gross income is about $60 per m2*. It is obvious that the phytomonitor also helps to improve climate control setting and optimize crop treatments so that the added value of the phytomonitoring application may cover initial investments during the first season. The phytomonitoring application is especially beneficial in warm areas where a grower has a routine task to combat plant stress and has to supervise climate and irrigation control with great care.

6. *Is there any difference between greenhouse and open field applications?*

Advantages of the phytomonitoring become more apparent in intensive cultivation technologies. Many of greenhouses are well-equipped growth facilities while irrigation is mostly only operable factor in the open. Nevertheless, there are some extremely promising phytomonitoring application in vineyards and orchards that have leaded to designing of special heavy-duty phytomonitoring stations equipped with solar power supplies and RF or cellular communication accessories.

7. *Are there any recommendations for using the phytomonitor in scientific researches?*

The LPS-03MA phytomonitor is a perfect tool for investigations in plant physiology and related sciences where the long-term evolution of plants and/or comparison of plant short- and long-term responses are the research objectives.

———*Meijaard D.,1995.The greenhouse industry in the Netherlands. In: Greenhouse Climate Control: An Integrated Approach. Wageningen Pers. Wageningen. P.11

8. *Can the Standard Phytomonitor be supplemented with the photosynthesis monitoring unit?*

At the moment the PhyTech Ltd. produces PM-48 Photosynthesis Monitor, which was specially designed for long-term monitoring of the leaves CO_2 exchange and it is fully compatible with up-to-date Standard Phytomonitor.

9

The Decision Support System for Agro-Technology Transfer

INTRODUCTION

The Decision Support System for Agrotechnology Transfer (DSSAT) is a software package integrating the effects of soil, crop phenotype, weather and management options that allows users to ask "what if" questions and simulate results by conducting, in minutes on a desktop computer, experiments which would consume a significant part of an agronomist's career. It has been in use for more than 15 years by researchers in over 100 countries.

DSSAT is a microcomputer software product that combines crop, soil and weather data bases into standard formats for access by crop models and application programmes. The user can then simulate multi-year outcomes of crop management strategies for different crops at any location in the world.

DSSAT also provides for validation of crop model outputs; thus allowing users to compare simulated outcomes with observed results. Crop model validation is accomplished by inputting the user's minimum data, running the model, and comparing outputs. By simulating probable outcomes of crop management strategies, DSSAT offers users information with which to rapidly appraise new crops, products, and practices for adoption.

The release of DSSAT Version 4 incorporates changes to both the structure of the crop models and the interface to the models and associated analysis and utility programmes. The DSSAT package incorporates models of 27 different crops with new tools that facilitate the creation and management of experimental, soil, and weather data files. DSSAT v4 includes improved application programmes for seasonal and sequence analyses that assess the economic risks and environmental impacts associated with irrigation, fertiliser and nutrient management, climate change, soil carbon sequestration, climate variability and precision management.

DSSAT is one of the principal products developed by the International Benchmark Sites Network for Agrotechnology Transfer (IBSNAT) project supported by the U.S. Agency for International Development from 1983 to 1993. It has subsequently continued to be developed through collaboration among scientists from the University of Florida, the University of Georgia, University of Guelph, University of Hawaii, the International Center for Soil Fertility and Agricultural Development, Iowa State University and other scientists associated with ICASA.

MINIMUM DATA

The minimum data set (MDS) refers to a minimum set of data required to run the crop models and validate the outputs. Validation requires:

1. Site weather data for the duration of the growing season;
2. Site soil data; and
3. Management and observed data from an experiment.

1. MDS Weather Data

The required minimum weather data includes:

- latitude and longitude of the weather station;
- daily values of incoming solar radiation (MJ/m^2-day);
- maximum and minimum air temperature (°C); and
- rainfall (mm).

Accessory data sets, such as daily dry and wet bulb temperatures and wind speed, are optional.

The period of weather records for validation must, at a minimum, cover the duration of the experiment and preferably should begin a few weeks before planting and continue a few weeks after harvest so that "what-if" type analyses may be performed as desirable.

2. **MDS Soil Data**

Desired soil data includes soil classification (e.g. USDA/NRCS), surface slope, soil color, permeability, and drainage class. Soil profile data by soil horizons include:

- upper and lower horizon depths (cm);
- percentage sand, silt, and clay centent;
- 1/3 bar bulk density;
- organic carbon;
- pH in water;
- aluminum saturation; and
- information on abundance of roots.

3. **Management and Experiment/Observed Data**

Management data includes information on planting date, dates when soil conditions were measured prior to planting, planting density, row spacing, planting depth, crop variety, irrigation, and fertiliser practices. This data are needed for both model validation and strategy evaluation.

In addition to site soil and weather data, experimental data includes crop growth data, soil water and fertility measurements. This data are needed for model validation.

PRODUCT INFORMATION

System Requirements:

- Personal Computer running MS-Windows
- Pentium2 with 128MB RAM minimal

- Pentium4 with 512MB RAM recommended
- CD-Rom Drive
- Hard Disk Space requirement 300MB of free space

Components

The Cropping System Model (CSM) released with DSSAT Version 4 represents a major departure from previously released crop models in DSSAT, not in function, but in design. The computer source code for the model was restructured into a modular format in which components separate along lines of scientific discipline and are structured to allow easy replacement or addition of modules. CSM now incorporates all crops as modules using a single soil module and a single weather module. The new cropping system model now contains models of 17 crops derived from the old DSSAT CROPGRO and CERES models. The major modules are:

- Land Module
- Management Module
- Soil module—A soil water balance sub-module and two soil nitrogen/organic matter modules.
- Weather module—Reads or generates daily weather data
- Soil-Plant-Atmosphere module—Deals with competition for light and water among the soil, plants, and atmosphere
- CROPGRO plant growth module:
    - Grain Legumes - Soybean, peanut, dry bean, chickpea, cowpea, velvet bean, and faba bean
    - Vegetables - Pepper, cabbage, tomato
    - Grasses – Bahia, brachiaria
- CERES Plant Growth Module:
    - Grain Cereals—Rice, maize, millet, sorghum, wheat, and barley
- SUBSTOR Plant Growth Module
    - Potato

SOIL ORGANIC MATTER (SOM)

A SOM–residue module from the CENTURY model was incorporated in the DSSAT crop simulation models, and a residue layer was added on top of the soil. Modifications were also made in the senescence module of CROPGRO, a model within DSSAT, so that senesced material is now added daily to the soil. By incorporating the CENTURY SOM–residue module, DSSAT crop simulation models have become more suitable for simulating low-input systems and conducting long-term sustainability analyses.

DSSAT version 4 allows consistent access to the crop models, data, input and output tools, and analysis programmes. The hierarchy is commodity-based within a tree structure where model inputs can be created and results analysed. The basic menu structure of DSSAT version 3.5 is maintained for backward compatibility.

A suite of tools is supplied for data management and analysis. XBuild is used to create and modify experiment files (X-Files). The suite of tools includes (but is not limited to) ATCreate (observed data), Weatherman (Weather data), GBuild (Graphing of outputs), and SBuild (Soil database).

In addition to the suite of applications installed with DSSAT, a number of accessory tools can be installed. These tools (ICSim, Stats, EZ Grapher and others) are applications that access the data in DSSAT for applications designed by the developers. The functionality exists for users to dynamically add their own application to the DSSAT toolbar.

10

Agro-Imperialism

Green Revolution to Biotechnology

SUPER IRRIIS THE GREEN REVOLUTION

Super IRRIIs the Green Revolution (GR) a curse or blessing for us? Four decades after the introduction of the GR, the time has now come to evaluate its overall impact. It seemed that political imperialism bade farewell to the third world countries, but in reality it exists in different forms. Agro-technology is such a form that has invaded the developing countries in the name of development. Under the guise of technology transfer, the neo-colonialists have entered into the developing countries, capturing their economy, culture, and society.

The technological transformation in agriculture is termed as the GR. It was introduced in our country in the 1960s. 'Miracle Seeds' are in central to the GR. Nobel Laureate agriculture scientist Norman Borlaug invented 'miracle seeds' through bringing out dwarf hybrid wheat seeds in 1954.

It is necessary to understand the geo-political and social condition of the 1950s to understand agro-imperialism. In the middle of the 20th century, colonial powers started withdrawing from their colonies in Asia, Africa, and Latin America, giving independence to various countries including India, Pakistan, Indonesia, Malaysia, and Vietnam.

Resentment was brewing among the peasant societies in these newborn countries for land and agri-reforms. The Communist Party of China, which came to state power at that time, extended its support to the peasant societies of these countries. The direct patronisation of the Communist Party of China to the peasant bodies became a serious concern for the US and its western allies. The US, which was looking for new strategy to exert its influence on these countries, used this miracle seed of Norman Borlaug as a magic medicine to thwart China.

The Rockefeller and Ford Foundations, which played a pioneering role in establishing the International Wheat Research Institute in Mexico, now came forward to transfer the hybrid seeds in Asia through setting up the International Rice Research Institute (IRRI) in the Philippines in 1960.

In the meantime, under the financial assistance of these two organisations, agriculture scholars from India, Pakistan, Indonesia, and the Philippines went to the USA for higher studies and training in US-agriculture.

After having been brainwashed in technological agriculture production, they returned back home and became desperate to dismantle the thousand years old traditional agriculture system, ignoring the people, soil, climate, land ownership system, nature and bio-diversity of their own countries, and terming the traditional system as 'outdated'

There was no dearth of money as USAID, the World Bank and other US-controlled organisations were ready to finance their ventures. The brainwashed scientists gave plans, and the politicians, already perplexed as to how to maintain political stability, grasped their ideas.

The dwarf hybrid rice seeds from IRRI came to the market in 1966. Under the direct patronisation of the government, the seeds quickly spread all over country with an ambitious pledge to uproot hunger, poverty and conflict from society. The crop production process engendered by this dwarf hybrid seed is termed as the Green Revolution.

The hybrid seeds, which were termed as High Yielding Variety or HYV, are in the real sense not high yielding. We could, more accurately, call them 'High Responsive Variety' as these seeds have the capacity to withstand high doses of chemical fertiliser, pesticide, and water. Not only do they withstand high doses, without application of high doses of inputs, not a single grain will come from hybrid seeds.

On the other hand, the traditional seeds cannot withstand high doses of these inputs. As a result, the western countries arranged all means to give with one hand and take away with other. They sold fertiliser, pesticide, irrigation appliances, etc. We had to set up fertilizer factories, but the machinery, technology, and technologists were theirs. That's not the end, fuel being used in irrigation is imported from foreign countries. We have to use our precious natural gas for producing fertiliser and electricity. A sizable quantity of electricity is being used for irrigation purpose. The power plant machinery was also imported from their countries. If you calculate the multiplier effects of all these inputs being used in cultivating HYV, you would find that the benefit is zero.

Farmers have slowly been added to the number of the landless.

What has the GR given to us? Has prosperity and peace resulted? Has poverty reduced? Has per capita income in real terms increased? Has per head food and nutrition intake increased? Statistics shows that there were no significant changes of these indexes. Rather, during the last four decades the marginal farmers have slowly been added to the number of the landless. Per capita income in real terms has not increased, the number of poor people not decreased, the gap between rich and poor widened, social inequality increased, and conflict, crime, and social instability multiplied.

The black-marketers, profit-mongering traders, and corrupt government officials control the supply and availability of seeds.

The GR presented our farmers dependency instead of making them self-sufficient. Our farmers had never been dependent on others for agri-inputs like seeds, fertilizer, and irrigation water. The found these in nature. The seeds of traditional indigenous varieties were under the full control of the farmers. The housewives used to preserve the seeds for the next season braving all odds. If somebody did not have, they easily got those from neighbours. But the hybrid seeds are no longer under the control of the farmers.

Seeds have now turned into consumer items. They are produced in specialised farms. Farmers, irrespective of their size, are at the mercy of the seed producing agencies. They have to pay huge sums of money for seeds. As a result, there is massive corruption before every sowing season. The black-marketers, profit-mongering traders, and corrupt government officials control the supply and availability of seeds.

HUNDREDS OF THOUSANDS OF CROP VARIETIES DISAPPEARED

The worst impact of the GR is that hundreds of thousands of crop varieties disappeared after the introduction of the monoculture system. Its hybrid seeds destroyed our rich genetic bio-diversity and natural and ecological balance.

Not only that — the over extraction of surface and ground water caused reduction in the water level. Most of the water bodies, including rivers, hoars, and beels, were silted up, turning them into arable land. As a result, water conservation capacity of our water bodies severely declined. Those, which hold a small quantity of water, have become unfit human use due to excessive use of chemical fertiliser and pesticide in the crop fields. On the other hand, although production of paddy increased to some extent, production of straw severely decreased.

Paddy straw is one of the main fodders in our country. The straw produced from hybrid seeds is not palatable for the cattle heads due to use of chemical fertiliser and pesticide. The impact is easily visible. The country has become cattle-less. Insects, earthworms, and frogs disappeared, resulting in decrease in poultry bird production. In a nutshell, the GR has totally destroyed our rich bio-diversity.

ALTERNATIVE TO GR

Was there any alternative to the GR? How would 140 million people of this small country be fed? The alternative can be found in nature. Nothing can be derived from nature by making it hostile. The exotic varieties failed to yield any benefit for our millions of marginal farmers. We have to go back to our roots. The success lies in the improvement of our indigenous varieties, development of cropping patterns, and proper use of resources.

But it's matter of great regret that our agricultural scientists do not do research with our own subjects, the teachers do not teach domestic issues. The agricultural institutions totally depend on western technology and books. The scientists-teachers run after lucrative scholarship offers in the west and having completed their courses, they come back brainwashed. They praise western agriculture and want to introduce western agriculture in our country in toto. But, they forget the basic difference between the society, land ownership pattern, and farmers of the west, and ours.

The American and European agriculture is farm-based. Their farmers (it's better to call them agri-industrialists) produce crops for industries. Most of our farmers produce crops for their own consumption. Our agriculture is labour-intensive, theirs mechanised. Think, what would happen if our total agriculture was mechanised? Millions of agri-labourers would become jobless. And it's happening. The rich farmers in northern Bangladesh are introducing tractors and other mechanical appliances in their farming, making thousands of day labourers jobless. They are moving towards towns and cities in search of work, swelling the slums there.

AGRO-IMPERIALISM

The agro-imperialism is not limited to the GR. With the passage of time, farmers of many third world countries are realising the hollowness of the GR. Farmers of many countries, including India and Southeast Asia, are returning back to the pre-GR traditional farming system. Organic farming is getting popularity fast. Now, the western multinational companies (MNC) have started propaganda, saying that the hybrid seeds failed to cope with the growing demand of food. It is the time to replace it with 'Genetically Engineered' or Biotechnology seeds. Our agricultural scientists have already swallowed the bait of the MNCs like Monsanto, Novartis, Dupont, Ciba Giegy, etc.

Our agriculture scientists are heading towards another technological disaster. Biotechnology seeds are even more dangerous than the GR hybrid seeds. The farmers have no control over hybrid seeds, but at least government agencies have some control. In contrast, control over biotechnology seeds will entirely be in the hands of MNCs. The list of the bad effects of GM food is long. Without mentioning them here, I only raise the issue of the risk to our farmers. Once the biotechnology seeds capture the market, the remaining genetic diversity will be quickly destroyed. Millions of marginal farmers of our country will then become pawns of the MNCs. They have to use seeds, fertiliser, and pesticide of a particular company. There would be no scope for bargaining as the government would have no control over those. But who will make our scientists understand? They have already sold themselves to the MNCs. The roots of agro-imperialism run deep.

Creating awareness is now the only means to avert the disaster. We have to make our farmers conscious; we have to build a social movement. In India, such a movement has already been started with the initiative of environmentalists and the media. Our sensible agricultural scientists and environmentalists should also come forward to combat the impending threat, otherwise a dark curtain will fall on our agriculture system and farmers.

11

Fertigation Systems and Chemigation Equipment

INTRODUCTION

Fertigation System and Chemigation equipment is a need for today's irrigation system. Optimum and efficient use of fertilisers is one of the major advantages of drip irrigation systems. We are well aware that success of micro irrigation system lies in precise application of fertilisers.

Fertigation system and Chemigation equipments are highly durable and composed of material inert to chemical and fertiliser reactions. Jain presents most reliable, accurate, automatic, energy efficient and suitable Fertigation Equipments to meet the exact fertiliser need of your crop.

Fertigation Systems and Chemigation Equipment Range includes:

FERTILIZER TANK

Features and Specifications:

- Mild steel construction.
- Powder coated with 70 to 80 micron thick deep blue coloured pure polyester from both inside and outside surface for protection against corrosion and weather effects.

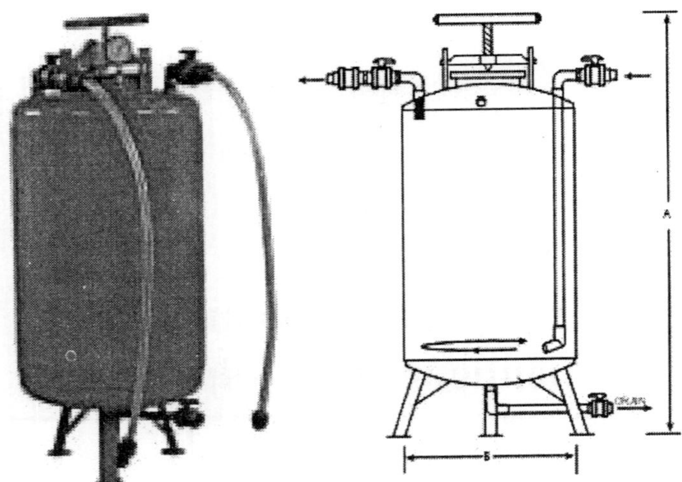

Fig. 11.1

- Turbulent inlet ensures thorough mixing of chemicals and/ or fertilisers.
- Separate valves are provided on the inlet and outlet to control the injection rate.
- Special provision of strainer on the outlet.
- Drain port to flush the tank.
- Maximum working pressure 10 kg/cm^2 (142 psi).
- Available in 30, 60, 90, 120 and 160 litre (8, 16, 24, 32 and 42 US gallon) capacity.

Applications:

- Fertiliser and chemical injection through drip and sprinkler irrigation systems.

FERTILISER INJECTOR

Features and Specifications:

- Mild steel construction.
- Powder coated with 70 to 80 micron thick deep blue coloured pure polyester from both inside and outside surface for protection against corrosion and weather effects.

- Specially designed high quality polyethylene accumulator maintains uniform injection rate.
- Fertiliser/chemical injection rate or proportion can be adjusted using flow control knob.
- Maintains the proportion constant even if the flow rate in the main line varies.

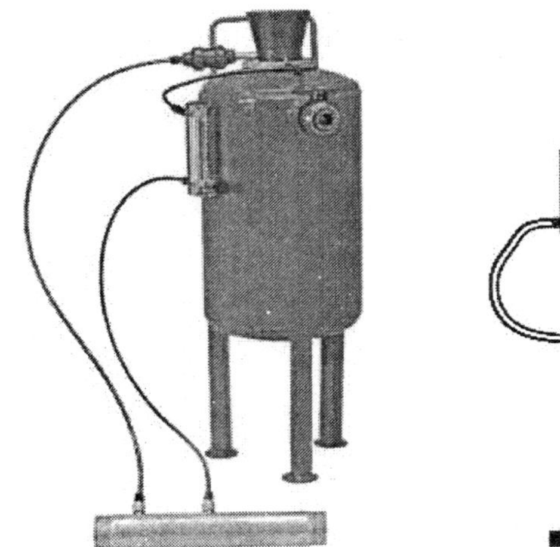

Fig. 11.2

- In-built ventury creates pressure difference between inlet and outlet thereby increasing the water pressure on the accumulator to inject the fertiliser.
- Flow meter indicates actual stock solution's injection rate.
- Available in 30, 60, 90, 120 and 160 litre (8, 16, 24, 32 and 42 US gallon) capacity.
- Maximum working pressure 10 kg/cm^2 (142 psi).
- Minimum flow to operate ventury 5 m^3/hr (22 gpm).
- Maximum capacity of the tank 30 litres.

Applications:

- Fertiliser and chemical injection through drip and sprinkler irrigation systems.

VENTURY INJECTORS

Features and Specifications:

- Made of engineering plastic.
- Excellent chemical resistance to most of the chemicals.
- Highly efficient and compact differential pressure injection device.
- Economical and low cost option.
- Available in ¾" BSP inlet/outlet connection.

Applications:

- Fertiliser and chemical injection through drip and sprinkler irrigation systems.

Different Installation options for Ventury Injector

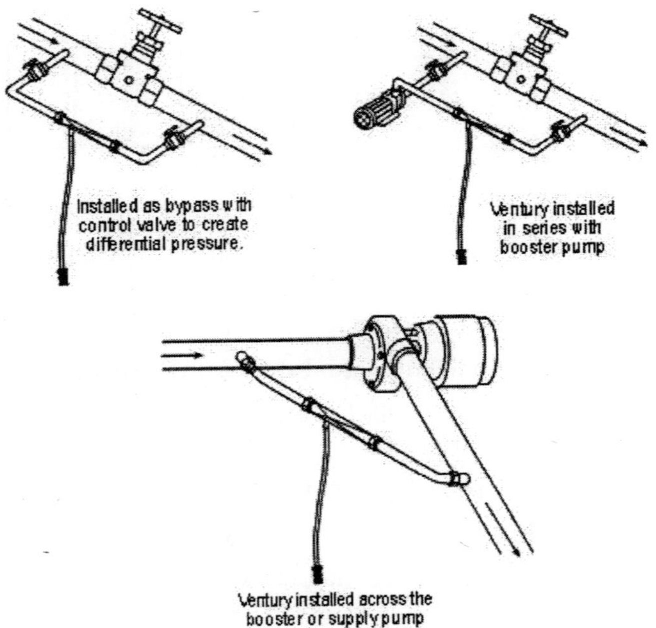

Fig. 11.3

J-INJECTOR PUMP

Features and Specifications:

- Compact design.
- Accurate and proportional injection rate.
- Non-electric, operates on hydraulic pressure.
- Dosing rate can be set externally to desired percentage.
- Low pressure loss.
- ¾" BSP inlet/outlet connection.
- Dosing flow range 1- 30 lph (0.26 - 7.94 gph).
- Side suction keeps the internal parts away from acids, chemicals, fertilisers.
- Self priming.
- Easy to connect and disconnect.
- Operating pressure range: 0.3 - 5 kg/cm^2 (4.27-71.11 psi).

Applications:

- Recommended to use where precise application of fertiliser and other agricultural chemicals are required.

MANIFOLD-FERTIGATION

Features and Specifications:

- Assemble in different options using metal and / or PVC fittings.
- Available in 1½", 2", 2½" and 3" BSP Female threaded connections.
- Any other size or end connection available.

Applications:

- Convenient and quick attachment of Ventury injector, fertiliser tank and injector pump to irrigation system.

12

The Approaches of Agricultural Development

INTRODUCTION

Since independence our country has seen many stages in agricultural development through different approaches in time-slots to develop resources for their optimum utilisation. Initially the focus was on human and community development, then to specific crops and fertilisers, treating the state as a unit of planning and lastly on development of resources and their optimum utilisation in a sustainable manner within the frame work of resource constraints and potentials of each region. In India food grain production more than quadrupled since the early 1950s from 51 million tons to over 209 million tons in 2000, while population nearly tripled from 350 million to one billion during the same period. Potentially high production areas (Eastern and Central States) are still lagging behind in productivity increases and in research there is a growing disparity between the actual and the potential yields which points to a crucial gap between research and extension.

The changing economic scenario in India and the need for appropriate agricultural technologies and agro-management practices to respond to food and nutritional security, poverty alleviation, diversifying market demand, export opportunities, and environmental concern is posing new challenges to the technology dissemination system.

SPECIALISATION AND SUSTAINABLE MANAGEMENT OF NATURAL RESOURCES

It is expected that future agricultural growth would largely accrue from improvements in productivity of diversified farming systems with regional specialisation and sustainable management of natural resources, especially land and water. Effective linkages of production system with marketing, agro-processing, and developing an army of trained Private Extension Service Providers, other value added activities would play an increasing role in the diversification of agriculture.

The past approaches of agricultural development were top-down in nature and had considerable success in resource rich and irrigated areas where profitability of new technology was high and management condition in farmers field were not very different as compared to that of research farms. However, these approaches resulted in limited success in rainfed areas due to large heterogeneity and complexity in farming conditions. It was also realised that farmers had a wealth of indigenous technological knowledge, which were being practised and upgraded through informal research inputs to meet the needs of changing agricultural scenario in rural areas as per need of climate, soil, inputs and economy. Thus within a broad agro-climatic region local conditions may result in several agro-eco systems, each with its own environmental conditions, however, similar agro-eco systems may develop on comparable soil landscape positions. The agro-ecosystem pronounced variations in climate is expressed in vegetation and reflected in soils.

THE AGRO-CLIMATIC REGIONAL PLANNING APPROACH

The agro-climatic regional planning approach was intended to take an integrated view of the agricultural economy in relation to resource base and linkages with other sectors implying future agricultural development specific to agro-eco regions with a multi disciplinary approach. Several key system constraints such as multiplicity of technology transfer system, narrow focus of agricultural extension system, lack of farmers' focus and feed back ,inadequate technical capacity within extension system lack of local capacity to validate and refine technology, limited research by KVK/ZRS, weak research extension linkage, poor communication capacity, inadequate operating resources for extension etc were identified.

It is becoming increasingly evident that public extension by itself can no longer respond to the multifarious demands of farming systems. There is need for reappraisal of the capacity of existing agricultural extension system to address, effectively, contemporary and future needs of the farming community. Public funding for sustaining the vast infrastructure is also under considerable strain. Meanwhile, in response to market demand the existing public extension network is inexorably being complemented supplemented and in some instances replaced by private extension.

Based on this fact, Agro-Ecological Situation (AES) approach evolved under National Agricultural Technology Project (NATP). This farmer's participatory approach is crucial not only to increase the relevance of a programme but also to make best use of available indigenous knowledge. The qualitative difference to other existing projects lies in the fact that NATP is based exclusively upon bottom-up approach. Under NATP, seven states covering 28 districts i.e. 4 districts from each state (Andhra Pradesh, Himachal Pradesh, Bihar, Jharkhand, Maharashtra, Orissa and Punjab) fairly representing each Agro-climatic Zone(ACZ), have been selected where Agricultural Technology Management Agency (ATMA) as an autonomous body at district level has been established for pilot testing of new institutional arrangements moving towards integrated extension delivery, adopting bottom up planning procedure, establishment of linkage among research-extension-farmer-NGO's/Corporate sector market, making the technology dissemination farmer driven and farmer accountable, ensure women participation in agriculture and create information connectivity to all blocks with district level ATMA with an ultimate aim of economic emancipation of farming community.

There are state government organisations in the district like DOA, DOH, DOAH, DOF, Cooperatives, Dairy development, Plant Protection, COMFED, DRDA, whereas the State Agricultural University has two of its colleges for Veterinary Sciences and Dairy Technology located in the district along with an Agricultural Research Station and a Krishi Vigyan Kendra in Patna and Barh respectively. These institutions are directly or indirectly involved in running a parallel extension system. In addition to Governmental organisations the district has a number of Non-Governmental Organisations (NGO's) registered, primarily involved in agricultural development. These forces would be organised and coordinated through ATMA, Patna under pilot testing of new institutional arrangement. ATMA Governing Board (AGB) and ATMA Management Committee (AMC) have been constituted with representation from different departments, farmers, livestock producers, horticultural farmers, women, scheduled caste farmers, banks, input suppliers, NGO's, Marketing Associations and fishermen farmers, at the district level. Similarly, at block level Farmers Advisory Committees (FAC) and Team of Farm Advisors (TOFA) have been constituted. At Panchayat and village level, Farmer Interest Groups (FIG), Commodity Associations (CA's), Self Help Groups (SHG's) and Farmers Organisations (FO's) are being organised to introduce a new form of extension delivery system which is decentralised, demand driven, farmer accountable, location specific, system based, participatory, well integrated with Research-Farmer-Market linkage and technically and financially sustainable in its orientation.

It is planned to involve persons responsible for extension delivery either directly or indirectly in capacity building towards the new technologies so as to meet not only the challenges emerging out of globalisation but also to explore newer and novel ways of developing the agricultural sector in general and farmers economic well-being in particular. This is proposed to be done through exposure visits, training in information technology to all the stakeholders through R.A.U., Bihar and MANAGE, Hyderabad, workshops and interface at district level. The members of FAC/FIG/FO/SHG would be linked with management of marketing boards locally known as Bazaar Samiti to minimise the role of middlemen and brokers and to make remunerative prices available. Better communication and stay facilities to the farmers. The general and common requirements of the farmers would be met by developing unemployed youth in the villages as Para-technicians by helping them develop their skills in areas of seed production, artificial insemination, graft and gootee preparation, fish-fry production, plant protection, plant protection equipment handlers, etc. They can work through Agri-clinics at reasonable user fee. Similarly, some farmers can be encouraged and developed for diversifying in allied agricultural activities like bee-keeping, mushroom production, medicinal and aromatic plants, floriculture, Agri-horti systems, rural godowns and marketing etc. Finally, success stories from different fields of agriculture emerging out of the working arrangements and endeavours would be identified and documented for their replication in other similar conditions of the district.

This chapter has emerged through participatory rural appraisal from all the five agro-ecological situations of the district by selected and trained research and extension officers of the district. The programmes proposed to be undertaken are demand driven, agro-eco situation specific, and multi-disciplinary in system envisaging bottom-up approach with practical implications. This bottom-up approach cares not only for the production system but also about the other stakeholders of the farming system like market information about demand and supply, input, credit institutions, etc. who were earlier left out when top-down approach was in vogue. Sustainability of bottom-up planning would however, depend not only on the active interest of bureaucracy in general and Chairman of ATMA in particular but also on the general agreement of all the stakeholders who have the benefit of farming community in the core of their hearts.

According to some experts higher degree of efficiency, quick decision, starting of activities without delay are some of the positive aspects of Planning from Top. They also see certain disadvantages in the participatory planning process like it is time consuming, discussion and decision-making process within large and heterogeneous group on different levels is difficult and complicated. The negative side of planning from above is low social acceptance as transparency is not ensured and people are less willing to participate in the implementation process. The long tradition of development planning from 'outside' and from 'above' is in itself not too impressive. People's involvement in planning is now being recognised by all as essential and must for rural development. The true essence of Participatory Planning and Participatory Monitoring and Evaluation (PM&E), lies in total involvement and decision-making by people at every stage of Project Formation, Implementation, Monitoring and Evaluation.

13

Advanced Organic Fertigation

INTRODUCTION

A tailor-made package of technology and know-how is now available to growers of organic crops by one of the world's largest supplier of irrigation equipment and know-how.

Low volume irrigation—a relatively new technology—which was initially developed to save water, has found even further applications in bioorganic agriculture, in particular the efficient application of liquid organic fertilisers and compost making.

The key to bio-organic agriculture is to establish optimal plant nutrition and production while restoring the soil and its micro-organisms to its natural state. The use of inorganic fertiliser, pesticides and herbicides are forbidden, therefore compost and other organic nutrient sources play a key role.

However, the use of natural nitrogen sources for fertilisation through the irrigation system needs special attention. For example, proper filtration, maintenance and periodical flushing of the drip system in order to avoid possible clogging are necessary.

Oscar Lutenberg, head of Netafim's bio-organic division said that the company has developed a free-flow fertigation system that operates for a long time.

"without any malfunctions."

One of the major problems that affect the efficiency of irrigation in organic agriculture is the creation of bacterial slime and lime scale, which can decrease flow rate and even clog the dripper. This problem has been overcome by providing suitable disinfestation that enables the free flow of nutrient-enriched water through the pipes.

"We use only materials approved by the IBOAA and the Plant Protection and Inspection Service, such as Sodium Hypochlorite and several organic acids for breaking down the scale layer in the inner tube," he said.

COMPOSTING

Composting is vital in organic production, efficiently treats and stabilises fresh organic material, destroys weed seeds and pathogenic microbes, reduces the volume of organic waste and prevents environmental pollution. Adding compost to the soil improves its physical characteristics, aerates heavy and clay soils, improves the water retention capacity of light and sandy soils, stimulates the development of biomass, reduces the carbon/nitrogen ratio and provides the soil with Humic elements.

When used in compost preparation, low-volume irrigation technology—whether micro-sprinklers or drip-lines—is very effective in preventing the leaching of nutrients and consequently environmental pollution.

POTENTIAL

Netafim, the Israeli company which pioneered drip irrigation in the 1960's, is today one of the world's largest companies manufacturing and marketing irrigation systems. It saw the potential for its systems in organic production some three years ago when it became an active member of the Israel Bio-Organic Agriculture Association (IBOAA) and last March of the international organic agriculture association (IFOAM).

Its entry into the organic market began in Greece, where the company worked with farmers growing citrus, olives, and corn, the latter slated for feeding chickens in the production of organic meat.

Since then it has developed a "Bio-Organic Package," which it launched earlier this year. As well as recommendations for irrigation equipment, the package includes technical guidance to provide the grower with the ways and means to maintain an optimal organic system.

THE BIO-ORGANIC PROGRAMME

- Hydraulic planning and solutions to fit the requirements of each and every farmer, taking into account the environmental conditions, topography, different soil types and the water sources available. Systems are flexible and are designed to produce maximum yield at minimum cost.
- Deployment of an accurate irrigation system that controls weeds and can carry different water qualities.
- The supply of organic oriented and approved pest management products such as fruit and olive flytraps and the use of bees and wasps to eradicate pests.
- Comprehensive support for irrigation and nutrient management for all crop types in open field and greenhouse cultivation; compost application; equipment maintenance and disinfestation.
- The application of an irrigation system that fits itself to the type of crop; these can be either an above-surface or subsurface drip irrigation system.
- In fruit plantations, Netafim uses "feeding channels" through which the trees are provided with a three-year compost supply. This technology uses open ditches with above-surface drippers and, when necessary, micro-sprinklers. Netafim has pioneered this technology especially for organic cultivation.
- In open field cultivation, the emphasis is on water-saving using accurate placing of the dripper near the target which controls weeds and prevents the leaves from getting wet;
- Advise on the accurate spread of compost and installation of dripper lines on top of the compost in order to provide the crop with all the nutrients it requires through the drip irrigation system.

CONVENTIONAL

Low volume drip irrigation systems employed in the bio-organic systems were originally designed to save water. They originated in Israel's Negev dessert, where dry conditions and a limited water supply created the need for an environmentally friendly crop watering system.

Today they allow large-scale delivery of precisely controlled amounts of water and fertiliser directly at the plant's roots. Cost-effective and easily deployed, they eliminate water waste and increase crop yield. Hence conventional growers through Netafim's Turnkey Agricultural Projects (NETAP) can just as easily use these systems and even the bio-organic package.

The reason for developing the system further and entering the organic niche was "the global consumers' trend for safe food, which is presumed guaranteed by practicing the organic agriculture principles, while maintaining soil-plant-water balance and preserving the ecosystem," says Lutenberg.

"What guides us these days in organic agriculture is our perception that organic agriculture is practicing the ancient agricultural principles using the technology and know-how of the present for a better future to mankind and the environment," he said.

BIOFUEL CROPS AND BIOFUEL ENERGY

The Biofuels industry will need to rely on advanced agricultural practices that will integrate the multidisciplinary fields of genetics, agronomic and economical evaluations.

Most of the "candidate crops" for energy production can be cultivated under irrigated or rain-fed conditions. However, yield and production of most crops (maize, sugarcane, sugar beet and others) is nearly doubled under an optimal irrigation regime.

In cases where unstable climate conditions and low rainfall are expected, low yield will result, causing irregular raw material supplies and deficits for industry.

The Biofuels industry will have to ensure a constant supply of raw materials with uniform quality parameters, while growers will need to base production on a reliable water supply.

Producers that focus on assuring that the right quantity and quality of plant raw materials will undoubtedly achieve commercial success.

It is a well-known fact that the threat of water shortages is growing in various parts of the globe. Furthermore, in many locations, water availability is challenged by problems of infrastructure, transportation and water storage.

The future of "Biofuels crops" in industry will depend on a regular water supply where environmental protection is balanced with improved agricultural practices.

The Environmental Drip Irrigation platform is based on the uniform application to the active root system zone, of precise amounts of water and fertilisers, combined with rational agricultural knowledge of the plant's water needs.

Environmental Drip Irrigation is directly linked to the reduction of pesticide application and optimal use of fertilisers. Drip irrigation teamed with the rational application of pesticides prevents fertilisers from accumulating in the soil and in groundwater. Using monitoring and models for predicting disease development and understanding plant mineral demands—are substantial factors in minimising environmental damage.

Environmental Drip Irrigation is extremely economical both in saving water and preventing dissipation of water resources. Furthermore, it is an effective solution against soil erosion, and also significantly reduces energy usage compared with most other irrigation methods.

In conclusion, Environmental Drip Irrigation integrates agricultural and ecological concerns at their most optimised levels. The Biofuels industry, by adopting the Environmental Irrigation concept, would be a great deal more effective in meeting two all-important goals: to assure regular supplies of raw materials needed by the industry and to help correct growing imbalances in our environment.

14

Conventional Irrigation of Grapevines

INTRODUCTION

Over 2000 years of tradition, viticulture (growing wine grapes) has been adapted to changing needs and tastes. The current move to advanced irrigation in traditional grape production regions reflects a need to compete with consistent yield quality in a global market, and to deal with world climatic changes.

RESTRICTED DEFICIT IRRIGATION FOR VINEYARDS

Controlling irrigation application often involves supplying less water than required for optimal vine development. Known as Restricted Deficit Irrigation (RDI), this practice is designed to keep the vine under controlled stress conditions, to increase wine grape quality. RDI gives grape growers control over:

- Berry size
- Vegetation and light regime
- Availability of assimilates, photosynthesis
- Turgor, solids concentration

Effective application of an RDI strategy is contingent upon stringent and accurate irrigation control, possible only through use of advanced irrigation techniques like drip irrigation.

When creating RDI strategy, grape growers need to take into account numerous factors, including the vine canopy, soil resources, and climatic conditions of both the previous winter and current season. Specifically, environmental conditions such as rainfall (quantity and timing), water retention, and evapo-transpiration should be analysed.

Before rolling out an RDI strategy, the following questions need to be specifically addressed:

1. When to start? Irrigation usually commences a certain period after the end of the rainy season, and a number of factors will determine when it is optimal to begin irrigating.
2. How to start? How to best irrigate while the canopy is being built and fruit sets?
3. Irrigation frequency? How to most effectively irrigate from the veraison to the end of sugar build-up? How should irrigation be modified after maturity?
4. Irrigation after harvest? How to continue irrigation most effectively after harvest, if rains haven't begun?

FREQUENCY OF IRRIGATION

Unlike conventional irrigation methods, drip Irrigation systems does not consist of utilising the soil as a water reservoir, since the horizontal water distribution is limited and depends on the hydraulic conductivity properties of the soil. Therefore, this method requires frequent irrigations in order to eliminate water percolation to soil layers bellow the major root zone. The frequency is determined by three major factors; soil properties, evaportranspiration rate and water emission rate of the irrigation system. Drip systems are normally designed to supply the water requirement of one day at least at the top season. At the beginning and the end of the season when the water requirements are low due to low evaporative demands and low canopy surface irrigation frequencies may vary between one and two per week whereas in during the major season frequencies may increase to daily or every other day. Sandy soils require more frequent irrigations than heavy soils. Accumulation of large deficits may require extended duration of irrigation hours and may result in water percolation or run off.

Tensiometers or other soil moisture sensors located at about 40 cm horizontal and vertical distance from the drippers may serve as useful tools for adjusting the rate of irrigation besides the use of meteorological data. Constant soil water readings indicate proper irrigation coinciding with the rate of consumption. Increasing or decreasing weekly readings indicate excessive or deficient irrigation respectively. In case of needs for water stress applications during various stages of growth and development along the season, reducing the amount of water per application rather than increasing the intervals is preferable. This way provides better control of the stress severity.

Irrigation control by soil sensors can also be done automated by the use of irrigation computer which is capable of averaging the readings of a few sensors and turn on and off the irrigation system at pre determined threshold values.

In addition to soil water sensor for the determination of the threshold for irrigation it is also advisable to place a sensor at a depth of 1.2m in order to monitor possible percolation. Water percolation to layers bellow the main root zone is undesirable since it causes miscalculation of the amounts of water and minerals supplied by fertigation intended for consumption by the vines. Percolation of minerals particularly nitrogen may contaminate the ground water and cause environmental damage.

TABLE GRAPES IRRIGATION

Modern table grapes vineyards require more intensive irrigation than wine grapes due to the use of training systems designed to accommodate large leaf area for higher production. The quality requirements for table grapes are different than those of wine grapes since appearance rather flavour and aroma is mostly desired. Contrary to winegrapes, large berry size is a major objective in this case and maintenance of high during the period of berry development is essential in this case. The maximal recommended pan evaporation coefficients in intensive table grapes vineyards is 0.6-0.7 and the duration of such high irrigation rates vary according to the varieties. It is shorter in early ripening varieties and longer in the late varieties. High crop levels cause increased water loss from the canopy, therefore higher coefficients should be used in years with high crop levels. The harvest time is sometimes determined by the market demands and fruit might be held on the vines for extended periods, beyond their physiological ripening stage.

Irrigation of early varieties after harvest is normally aimed at maintaining the foliage in a good physiological state and thereby enable the formation of reserve sugars for the next season and cane maturation. Excessive irrigation after harvest may stimulate undesired vegetative growth which normally does not undergo efficient fruit bud differentiation and incomplete can maturation which may not survive the winter temperatures and weaken the vine.

FERTIGATION PRINCIPLES

Application of soluble fertilisers through the irrigation water is essential in drip irrigation regimes since the efficient occupation of the limited soil volume under the dripper accelerates the rate of mineral depletion. On the other hand, fertigation provides an efficient means for controlling the concentration of minerals in the immediate vicinity of the active root system. Plant roots cannot read measures such as kg per hectars since the uptake of minerals is a function of mineral concentrations. Fertigation requires to look at the mineral concentrations in the irrigation water and in the irrigated soil volume in addition to the common measures of kg per hectare. The efficient uptake of minerals in drip regimes is due to the maintenance of biological concentrations at high water availabilities, good aeration and formation of concentration gradients. Such mineral concentration gradients expose various parts of the root system to a wide range of concentrations so that part of the root system is always exposed to optimal concentrations and uptake rates. The efficient uptake is reflected in the entire root system due to exchange of water and minerals between individual roots. The fast depletion of minerals is reflected in a rapid response of the vines to the cessation of application of elements such as nitrogen which is known to accelerate vegetative growth.

RATES, TIME AND AMOUNTS OF APPLICATIONS

For grapevines, the optimal concentrations of NPK in the irrigation water are 150, 20 and 150 ppm respectively. These concentrations are used for stimulating vegetative growth when needed while lower concentrations can be used when moderate growth is needed. It is not recommended to apply nitrogen after veraison since it may stimulate vegetative growth which is undesirable at this phonological stage due to assimilate competition between the growing tips and the ripening fruits. On the contrary, phosphorus and Potassium may be applied throughout the entire season since they do not inerfere with growth and are not as mobile as Nitrogen. Microelements can be added according to the needs, the uptake of iron chealates is particularly efficient in drip irrigation regimes.

The selection of the nutrient solution's composition is determined by the needs. Leaf mineral analysis on one hand and soil extractors on the other are very useful tools for this purpose. Petiole rather than blade analysis better reflects the nutritional status of the vines. Petiole sampling is normally done at full bloom, however veraison and preharvest are suitable dates as well. Recommended levels for petiole and blades for each sampling season are available in the literature. The advantage of full bloom sampling is the ability to apply the results at the same season. On the other hand, petiole nitrate nitrogen content is quite unstable at bloom time and does not always reflect the need for nitrogen fertilisation while pre harvest analysis is more reliable and together with the vigour performance of the same season may serve for the next season's application policy. In early table varieties it is still possible to apply nitrogen at the same season provided the vegetative growth is insufficient as well. Autumn nitrogen application is not recommended in drip irrigation regimes since the rates of water application in this season are rather low and the high leaching risk during the winter rains. Potassium needs are also better reflected at harvest time since it is the major element in the fruit and its content in the petiole is at its minimum at this time. High petiole potassium is well correlated with high must potassium which is known to increase wine pH. In regions where wine pH tends be high, it is it advisable to restrict the potassium application to the levels required for its physiological functions. However, in table grapes high potassium and high pH mask the sourness perception and may enable to harvest the grapes at lower sugar contents. Phosphorous increases fruit bud differentiation and sugar accumulation but tends to reduce berry size. These are very important properties for winegrapes but may be useful for table grapes according to the local conditions and the varieties.

THE USE OF SOIL EXTRACTORS

Soil extractors are used for monitoring the concentrations of various minerals in the root zone. Extraction can be done only when the soil is saturated and concentrations may be read by the use of field kits. The use of such extractors is important particularly to monitor excessive concentrations which may result from the difference between the rate of application and uptake.

15

Farm Management and Agri-Economics

INTRODUCTION

Our agricultural scene is a changing rapidly since the mid-sixties when we experienced the 'Green Revolution'. It was followed by the 'White Revolution', which brought in the flood of milk in the country. Now, we are on way to 'Yellow Revolution' i.e. revolution in fruit production. An important fact behind these revolutions is the economic consideration. Farmers have been adopting the new technology in agriculture and related fields mainly because of these innovations pay more all compared to their tradition crops. Agriculture has become more market oriented. The considerations of cost of production, income and profitability are summing significance. Thus, the Agri-economics has come in. The farm management is the field closely related to agri-ecomnomics. There is no doubt that farmers adopting new technology are quite conscious about the management of their farm. However, there is general apathy about the farm management amongst the farmers, Agri-scientists and the policy makers. Being a relatively young since, no much attentions is post to it. There may be some sound reasons for this. However, with changing scenario in agriculture, the need for guiding farmers in management of their farms assumes significance. In any business proposition, management has to play a significant role. In other sectors of the economy, 'management' has become a catchword. In recent years, the irritations giving training in 'management' are increasing by leaps and bounds. There is good demand for 'managers' trained in these institutions. Now, the technical persons having engineering and other vocational background are turning towards MBA and allied courses. However, with rare exceptions, there are no such Institutions for training persons in farm management. These are institution for guiding farmers in this subject one of the reasons for this may be that our farming is relatively a small-scale farming and the scattered one. However, looking to the changes that are taking place. In agriculture sector, there is need for guiding the farmers in this economic aspect. The subject matter covers several aspect such as decision-making, farm planning, resource planning (land, labour, capital), farm costing, farm accounting, marketing, analysis of case studies of successful farmers. The use of efficiency measures to evaluate the enterprise is also an important aspect of farm management. In the recent years, the statistical tools such a production function, linear programming are being used to get an idea about the resource. We in an enterprise and for getting an optimum plan for a given farm taking into account the resources and the limitations in which it is being operated. In the near future the farm management is likely to assume move and move significance.

DECISION-MAKING IN FARMING BUSINESS

Decision-making in any business largely depends on the nature of business. To understand decision-making in farming, it is necessary to know the nature of Indian farming.

(a) Nature of Indian Farming

1. Indian farming is basically a subsistence farming. But due to new technological developments in all the areas of production, productivity of crops and live-stock has increased substantially, total production has increased several times and marketable surpluses with the farmers have also increased. As a result farming has assumed commercial proposition.

(b) Indian farming is a diversified and mixed type of farming. The farmers grow 5-6 crops or even more on their farms.

1. to meet their family requirements
2. to suit their soils on the farms and considering availability of resources including irrigation and market facilities. They also follow mixed farming, that is they keep few milch animals and follow dairy and also undertake sheep or goat keeping poultry keeping, etc. to supplement incomes from crop production. Therefore decision-making on such farms is more difficult and complex than on specialised farming. Indian farming is a family type farming. Farm family provides major part of labour and also provides some capital.

Role of Farmers

Indian farmer plays following four rules simultaneously:

1. He is an entrepreneur
2. He is a Manager
3. He is a Financer
4. He is a Labourer or Worker, He takes decisions in two capacities as an entrepreneur and a manager.

Area of Decision-making

There are four major areas of decision-making:

1. Production of crops and live-stock
2. Acquiring inputs
3. Obtaining credit
4. Marketing his produce

THE DECISIONS RELATED TO PRODUCTION ACTIVITY

1. What to produce? i.e. selection of crops and live-stock activity. Whether to have dairy, goat keeping or poultry or their combination.

2. How to produce? Adoption of suitable technology; whether to continue old technology or use new technology or combination of both.

3. How much to produce? To consider family requirements in respect of foodgrain, vegetable, spices, fruits, milk, etc and also think of producing marketable surpluses. In that case, think of demand supply position availability of market and prevailing prices. This will also apply to production of cash/commercial crops.

INPUTS IN COMMERCIAL FARMING

In modern commercial farming, wide range of inputs are required. They include seeds of high yielding and hybrid varieties, fertilisers, pesticides, plant growth regulators (hormones), improved implements and machinery, etc. There are commercial and specialised agencies dealing in these inputs. The farmer has to take decision and make choices relating to:

1. From whom to buy?
2. At what price to buy?
3. How much to buy?
4. Whether to buy on cash or on credit?

Obtaining credit

Although the farm family supplies some finance, it is not adequate considering larger financial requirements of modern agriculture. Therefore, farmer has to borrow from outside sources. In this connection he has to take following decisions. There is risk in borrowing.

1. How much to borrow?
2. From whom to borrow? There are some alternative sources viz. Moneylenders, banks, government incentive schemes, friends, relatives, traders, etc. He has to make right choice.
3. At what rate to borrow?

4. What security/mortgage to offer?
5. How repayments to be made?
6. How to face risk situation?

MARKETING OF FARM PRODUCE

When surplus produce is obtained or created, its disposal or marketing advantageously becomes necessary. Following decisions are involved in the marketing of farm produce.

1. Where to sell? This is the most important decision because in rural areas big, wholesale and organised markets are limited. As far a individual farmer is concern, his produce is of small quantity, he has no transport facility, there are no good roads to go to distant markets. The farmer has to make choice within limited alternatives.
2. Whom to sell? There are various types of agencies such as villager trader, wholesale cum commission agents, co-operatives, government (for selected commodities). He has to make choice from among them.
3. When to sell? Immediately after harvest when pries are the lowest or wait for better prices. He is also constrained by cash requirements for family expenditure. He has to take decision under most unfavourable situation.
4. At what price to sell? Here the position of farmer is very vulnerable. He has almost no choice. Prices in organised markets (Regulated markets) are fixed by open auction and he has to accept the price quoted by the highest bidder even this price may not be remunerative. In unorganised markets, prices are fixed by traders and farmer is hardly consulted. He has very limited scope for decision-making.

Decision-making

1. Decision-making is an art as well as science. It is more a brain work. The quality of decision-making is dependent on level of formal education, knowledge of the subject and experience. It is a continuos process;

2. The responsibility of decisions is borne by the decision-maker in terms of rewards and punishments for taking right or wrong decisions. If he takes right decisions, production increases, income and profit increases, and his family welfare improves. This is a reward for him. If he takes wrong decisions, the level of production goes down, there is no profit or low profit and low family welfare. He gets self-punishment.

16

Degradation and Remediation of Irrigated Soils

DEGRADATION PROCESSES

Compaction of irrigated soils frequently occurs (See Table 16.1 on next page). Compaction means increasing of the bulk density of the upper horizons higher than optimal values. The value of optimal density depends upon the soil structure and for many agricultural plants varies from 1.0 up to 1.3 g/cm^3 more than 1.4 g/ cm^3 results in reduction of crops, difficulties in soils treatment and impossibility of their restration. More tight packing of aggregates under wheels' passages of heavy agricultural machinery, formation of soil crust and furrow pan are reasons of soil compaction. Thrombosis and closing of pores and cracks as a result of increased low-aggregated clay content and the appeared superdispersion of smectite minerals may lead to compaction as well.

Slitisation is a more complicated soil compaction than above mentioned one (up to 1.6-1.7 g/cm^3), with slikensides available and shifts of soil blocks relative to each other. These attributes are diagnostic criteria of slitisation and can be developed for 15-25 years. This can be explained by hydrophyllisation of colloids, formation of highly dispersed soil polymers and microaggregate cementation by soluble components Si, Mg, Al. Slitogenesis can be explained by the process of grating, i.e. mechanical disintegration of clay mineral cristallites which occur at critical density on the surface of blocks rubed at soil swelling.

Under dry condition these soils are characterised by extreme density, massive structure and cracking. Tenacity and plasticity are typical for humid condition. Methods for slitisation control are difficult and weakly developed.

Table 16.1: Indices and criteria of physical degradation and hydromophizm of irrigated soils

Indices	*Degrees of soil degradation*				
	Absent	*Slight*	*Moderate*	*Severe*	*Very severe*
Decrease of humus horison thickness %	<10	10-20	21-51	51-100	>100
Middle annual increase of eroded soils area %	<0.2	0.2-0.5	0.6-1.0	1.1-1.5	>1.5
Soil mass loss, t/year	<5	6-25	26-100	101-200	>200
Soil consolidation, g/cm³	1-1.3	1.3-1.4	1.41-1.5	1.51-1.6	>1.6
Water penitration, m/day Aggregates content, %	>1.0	0.3-0.1	0.1-0.3	0.01-0.1	<0.01
dry sieving, mm. >10	10-20	21-30	31-40	14-50	>50
10.0.25	60-80	50-59	40-49	30-39	<30
Critical groundwater table, m: mineral one	>5	3.1-5	2.1-3	2-1	<1
sweat one	>3	2.1-3	1.1-2	0.8-1	<0.8
increase more than critical one, %	0	<15	16-30	31-50	>50
Groundwater rise, m/year	0	<0.2	0.2-0.4	0.5-0.7	>0.7
Yield lost, %	<10	10-24	25-50	51-75	>75

Destructurization is most frequently appeared phenomenon at irrigation. It can be found first of all micromorphologically. In secondary-hydromorphic conditions, and at low level of cultivation simplification of microaggregates structure and increase of microzonality in construction of plasma material, decreases of porosity take place. The expansion of lumpiness together with dicreasing water resistance and the amount of agronomically valuable aggregates have been observed. Structure degradation is connected first of all with large intensity of artificial rain and duration of watering, longtime water standing on the soil surface, pressure of heavy agricultural machines and treatment of soils which are not of physically mature.

Dehumification. Humus condition in irrigated soils is determined by the level of agriculture and quality of irrigation waters. Al low level of irrigation agrotechnology humus stocks reduce by 10-30 per cent of initial value for 20-30 years (See Table 16.2 on next page). These are basically biochemical losses, at irrigation erosion large losses of humus stock are possible. In this condition the content of humic acids connected with calsium, is deacreased, while fulvic acids are increased. Chernozems as compared to the other steppe soils lose humus at irrigation with higher intensity, that is determined by low increases of crops, smaller area of perennial grasses and return of *olant residues in soil for humus neoformation. In secondary-hydromorphic conditions at* shallow mineral groundwater causing secondary soil salinity, changes of humus content and composition are more expressed.

Decarbonation. The presence of carbonates in top horizons promotes structure preservation and soils ability not to alkalite. At irrigation by sweet water the carbonate supply is saved. Irrigation by sodic water promotes $CaCO_3$ leachining from 0-1 m layer. At irrigation of sulfate-natrium water with mineralisation 3-4 g/l the forming soil solution is saturated on $CaCO_3$, that favours to preserve of soil carbonates.

The question about secondary salinisation and solonisation at irrigation is covered in the works of V.A.Kovda, B.G.Rozanov, V.V.Egorov, A.B.Novikova, N.G.Minashina, B.A.Zimovetz, F.I.Kozlovsky, I.P.Aidarov, E.M.Samoulova, N.P.Panova, E.I.Pankova , R.A.Baer, A.P.Buriukova, I.Szabolch, H.E.

Table 16.2: Indices and criteria of chemical degradation of irrigated soils

Indices		*Degrees of soil degradation*				
		Absent	*Slight*	*Moderate*	*Severe*	*Very severe*
NPP content decrease, multiple*		<1.2	1.2-1.5	1.6-2.0	2.1-5	>5
Decrease of humus stock at 20 yrs. in humus horison, relative %*		<10	10-20	21-40	41-80	>80
Danger pf solotzesity, % $CaCO_3$, content in 0-50 cm layer		<5	1.5-5	1.5-1	<1	0
Alkalinisation in layer 0-30 cm	HCO_3 -Ca, mg/eg	<8.0	8.8-8.5	8.6-8.9	9.1-9.5	>9.5
	pH water	<0.7	0.71-1.0	1.1-1.6	1.7-2.0	>2.0
Solotzesity in layer 0-30 sm	Absorbed Na in:					
	soil, % CEC	1 <1	1-3	4-6	7-10	>10
		2 <3	3-5	6-10	11-15	>15
	soil solution,	1 <0.5	0.5-3.5	3.6-7.5	7.6-10.5	>10.5
	aNa/aCaII	2 <1.0	1.0-5.0	5.1-10	11-15	>15
Absorbed Mg content, % CEC		<30	40-50	51-60	61-70	>70
Total toxic salts in Ap layer, %						
in soils with soda		<0.1	0.1-0.2	0.21-0.3	0.31-0.5	>0.5
in soils without soda		<0.1	0.1-0.25	0.26-0.5	0.51-0.8	>0.8
Increase of saline soils area, % in year		<0.5	0.5-1	1.1-2	2.1-5	>5
Decrease of soil redoz potential, mV*		<50	50-100	101-200	201-500	>500
Watersoluble fluorine, mg/kg		<6	6-10	11-20	21-25	>25
Nitrification ability, mg/kg		>30	20-30	10-19	5-10	<5
Nitrates cjntent, mg/kg		<40	40-60	60-80	80-100	>100

* - comparison with drylands (initial); CEC cation exchange capacity

1. - for soils with exhangeble Na content less 1%;
2. - for other soils.

Dregne and other researchers. At mineralisation of groundwater exceeding the critical level and absence of drainage secondary soil salinisation is developed. The critical level and mineralisation of groundwater change as the climatic zone changes. The more arid climate, the deeper groundwater table and the less mineralisation should be lest secondary soil salinization be developed at mineralisation of groundwater 3-5 g/l they will cause soil salinisation in steppe zone at the groundwater table 1.0, in dry steppe - 1.5, semidesert - 2.0 m. Agricultural plants have a different salt and solonets resistance.

The permissible concentration of salts in soil solution for salt-sensitive plants is less than 2 g/l, for salt resistant ones - up to 13 g/l (moderately saline soils). On saline soils at mineral groundwater tables closer 10m from the surface drainage construction is required to be conducted simultaneously with creation of irrigation system. In other cases the construction of a drainage is necessary to begin at the rise of mineral groundwater up to 3 m.

Mineralogical degradation. Intensification of the processes of destruction, disintegration and transformation of silt fraction minerals results in an increase of clay content. Probably also, that some part of clay minerals is also in the structure of strong units, which are destroyed under effect of irrigation. As a result of these processes the content of illites is most oftenly increased rare that one of labile minerals and chlorites in clay fraction.

At small reserve of easily weathering minerals loss of clay fraction occurs under effect of irrigation due to strengthening on of the processes of destruction and migration of clay minerals (See Table 16.3 on next page).

Magnesium chlorites and labile minerals of clay fraction are first to be destroyed the least tolerant weathering. At watering with sodic waters or at close groundwater table, containing natrium(especially soda), superdispersion of labile minerals takes place. This strengthens their migration and soil physical properties deterioration.

Table 16.3: Indices and criteria of mineralogical degradation of irrigated soils

Indices	*Degrees of soil degradation*				
	Absent	*Slight*	*Moderate*	*Severe*	*Very severe*
Smectite content decrease, %	<10	10-20	21-30	31-40	>40
Water peptised clay content, %	<1	1-5	6-10	11-15	>15
Appearance of clay mineral superdispersion	No		Yes		

On slopes irrigation soil erosion is developed. In order to prevent decrease of soil thickness the flow value is required not to exceed the rate of soil formation. In this case the washout of soil makes up 1 t/ha annully . Increase of infiltation properties with the help of cracking and artificial structureless soils mass reduce the surface flow at irrigation by 1,5-3 times, at washing off - by 2-10 times.

IRRIGATION BY SODIC WATERS

Irrigation by sodic waters causes processes of alcalization, solonetsesity, struture reorganisation, compaction, formation of surface crust, decrease of water infiltration rate, strengthening of humus and minerals migration. The similar processes are observed at irrigation by waters with mineralization 1-4 g/l (for chernozems more than 0,5 g/l) chloride-natrium composition, they are expressed to a lesser degree at irrigation by sulfate-natrium waters. The salt regime is of accumulative-pulse character: to the end of vegetative period accumulation of easily soluble salts is marked, which are leached partially or completely with precipitation by autumn-winter-spring period. At mineralization higher 3 g/l after 15-20 years of irrigation accumulations is made possible due to evaporations. In Russia 8 per cent of the land is irrigated by waters with mineralization higher 1 g/l, including — 5 per cent with concentration of 1-2, 3 per cent — more than 2 g/l.

The most effective method of simultaneous amelioration of such waters and soils is: gypsum application, the use of acid with carbonates, as well as applying of calcium fertilisers, delution by sweet water.

Simultaneous evaluation of different degrees of soil degradation is determined by a parameter or a number of parameters, possessing maximum degradation degree. It also needs to specify adverse processes, which have the hignest rate. Now-a-days soil on this parameter can have a small degradation degree, however because of accelerated deterioration soil may quickly become strongly degradated on this parameter.

REMEDIATION MEASURES

1. Rational water use (water rate up to 300 m^3/ha), watering of the layer which contains 85 per cent of plant roots in accordance with phenophase plant development; realisation of refreshing irrigation by the rate of 50-70 m^3 /ha.

2. Measures preventing or decreasing compaction effect of machinery: designing of new units with low pressure on soils and application of accessories to existing technique which reduce soil consolidation (expansion and doubling of tractors wheels decreasing of their pressure on soil, lesser exploitation of farm machinery increasing of capacity for fields travel due to the use of far-reaching and combined units with extension-type appliances, introduction of technological gauges and application of higher dozes of manure on them).

3. Erosion-preventive measures, first of all: reduction of a diameter and decrease of the rate of drops fall of artificial rain, decrease of eroded ability of water flows, secondly (increase of soil stability): preliminary irrigation with small norms, deep plowing, craking, trash conservation of a surface with straw (2 t/ha), destruction of surface crust, bedding (on slopes with up to 6_0), accommodation of plants in bands (on slopes with $6\text{-}12_0$).

4. Applications of different agrotechnical methods and whenever possible postponement of a part of early-spring processes, causing compaction, to autumn.

5. Application of gypsum, conditioners, lignin and other materials to improve structural stability and water penetration.

6. Measures for maintenance of positive humus balance: application of organic and mineral fertilisers, straw, peat, sapropels, keeping of maximum plant residues on fields, sowing of perennial grasses and stubble crops. Remediation measures at irrigation by mineral waters or saline soils and at the rise of salt groundwaters:
7. Drainage.
8. Washing.
9. Chemical ameliorations.
10. Growing of salt resistance plant crops and complete elimination of their biomass from fields. Different parameters of various soil degradation can be used for monitoring of irrigated soils and the development of remediation measures.

17

Agricultural Technology and Health

AGROTECHNICAL PROGRESS

Agrotechnical progress comprises research by farmers and public and private providers; invention, discovery or development of a technique; and adoption, from innovation by early users to diffusion by learning or extension. It includes everything from the development of basic agricultural tools to biotechnology.

Agrotechnical progress has repeatedly driven revolutions in food production and transformed human development, from the Neolithic settlement, as former hunter-gatherers became farmers, to the Green Revolution in Asia, which brought unprecedented rises in food production. Only since about 1750, however, has agrotechnical change been a main engine of steady human development, and only since the 1950s has it been deliberately harnessed toward such ends. Indeed, the irrigation and biochemical revolutions of the 1960s and 1970s, with all their imperfections, have led the world's greatest and fastest advance in human development.

AGROTECHNOLOGY AND HEALTH

Research, invention, and adoption of agrotechnology have played an important role in improving human nutrition and health. Agrotechnology has introduced more effective plant breeds (such as high-yielding varieties), enhanced land management techniques (such as terracing), and improved water management tools (such as irrigation). The adoption of these techniques has benefited nutrition, largely through boosting crop productivity, thereby providing employment and income to rural populations and increasing local and global food supplies.

Pro-poor agrotechnology produces results suitable for low-risk, profitable adoption in conditions faced by many smaller and more asset-deprived farms. Such technology offers long-term benefits to the poor by increasing labour demand, lowering risks, enhancing access to cheap, reliable sources of energy and micronutrients, improving water use efficiency, and helping poor rural communities to acquire key assets. Normally, such economic gains carry clear health benefits.

For example, cheaper and less variable micronutrients mean better immune function in times of higher disease incidence or work stress.

More assets provide collateral, so poor households can borrow to meet sudden health costs or food price rises. Better water use efficiency reduces the scarcity and distance of essential drinking water.

Increased labour demand improves health by raising hungry workers' income and thus their command over food. Policy choices may be needed, however, to minimise harmful side effects on health.

Just as agrotechnology can benefit health, good health can accelerate agrotechnical progress. Research and invention of agrotechnologies cannot benefit health unless farmers adopt them, and healthy farmers are likelier to seek out, afford, find, and try new technology.

AGROTECHNOLOGICAL HEALTH RISKS

Health gains from improved farm production and employment through improved income, nutrition, shelter, and access to water far outweigh negative health effects. But certain agrotechnologies can endanger health by affecting the natural environment, and the type and form of labour needed for agricultural production. These impacts should be anticipated (and health and agriculture policies coordinated) in order to identify effective ways of monitoring them, reducing the hazards, and developing treatments.

Tools and mechanisation. Physical injury in farmwork is a threat to agricultural productivity and worker health. The risks and effects differ depending on the technology used. A study in Bangladesh showed that 80 per cent of female users of modern threshing technology suffered pain in their waist and legs for a few hours after threshing, but 20 per cent of the farmers said that traditional threshing technologies had caused similar problems. Overall, investment in the new technology was felt to be worthwhile since it made the job easier.

Most physical injury incurred in agriculture is preventable. It is largely ignored, but probably causes more death, pain, and work loss (with much less offsetting output benefit) than agrochemicals and water resources development put together.

Water resources development. For families living near irrigation projects, this more convenient water source may reduce disease through cleaner water or greater availability, facilitating better hygiene. But if the irrigation water is contaminated, drinking it can spread infectious diseases such as cholera and lead to chemical poisoning through surface or groundwater transfer of agricultural and industrial chemicals. Stagnant water is also a breeding ground for disease vectors, especially mosquitoes.

Use of agrochemicals. Pesticide use in crop cultivation is often higher than optimal for profit maximisation, notably in rice cultivation. Farmers' unawareness of the dangers of agrochemicals, combined with poor regulation and enforcement, often leads to poisoning. In Ecuador, chronic dermatitis was twice as common among potato workers as among controls. Many agrochemical poisonings are due to suicide, but apart from that, agrochemicals cause millions of poisonings each year, the vast majority in developing countries.

Fertiliser use also affects health. Nitrates and nitrites from fertiliser are among the most common contaminants in drinking water; nitrate contact with mouth bacteria causes nitrate poisoning. Yet excess fertilisers are often used inefficiently. In China, only 30 per cent of fertiliser applications reach the crop; much of the rest ends up in water courses. In northern China, more than one-half of groundwater monitoring sites had nitrate levels above the allowable limit. Contamination aside, this harms health through inadequate water quantity and increases women's time and labour burden in finding alternative water sources.

More can and should be done to ensure safe and appropriate use of agrochemicals. Reducing pesticide use often improves health and usually cuts production costs. Likewise, when fertilisers get into drinking water instead of crops, both production costs and health suffer.

In Indonesia, biological controls (within integrated pest management) have greatly reduced pesticide applications, improved health, and raised farm incomes.

Plant breeding also matters in this context. Plant type improvement on poor people's farms is almost unambiguously good for human development, but must be selected to decelerate inappropriate use of agrochemicals. In China, India, and South Africa, where farmers had been forced to choose between low cotton yields (due mainly to bollworm) and increasingly massive pesticide applications, transgenic Bt cotton has had substantial health benefits and raised farm income, including that of poor smallholders.

THE EFFECTS OF HUMAN HEALTH ON AGRICULTURAL TECHNOLOGY

The impacts of human health on agrotechnology are complex and often mediated by the seasonal nature of both illness and labour demand. Temperature and rainfall determine survival and breeding patterns of mosquitoes, and thus incidence of malaria. The rainy season also sees a greater incidence of diarrhoeal diseases. Nutrition in pre-harvest seasons tends to be worse, increasing susceptibility to illness.

Such threats to human health often coincide with times of high seasonal labour requirements. This situation has implications for the use of agrotechnology because seasonal labour bottlenecks and illness during certain seasons can affect adoption of technology, either positively, as households improve technology out of necessity, or negatively, as households facing labour shortage and lower income due to illness are forced to spend resources on health care and have little left to invest in technologies that ease labour constraints. Where external inputs are used, money might be diverted away from these toward paying health care expenses.

Illness during the slack season is especially likely to deplete farm labour for long-run investments, such as conservation. Households made poorer by illness and needing to save seasonal labor are likely to target activities that give quick returns.

HIV/AIDS illustrates how disease affects agrotechnology. HIV/AIDS-related expenditures can reduce farm households' spending on productivity-enhancing inputs, especially given that the labour used to apply such inputs might not be available owing to death and time spent caring for the sick and attending funerals. HIV/AIDS also affects the relationship between labour and technology. HIV/AIDSstricken rural households often invest in labour-substituting technology, which is less likely to be affected by ill health than labour-intensive technology. This situation stimulates labour-saving technology, in particular long-lasting machinery such as tractors. This stimulation is perverse: it worsens poverty by skewing technical progress in ways that reduce demand for labour and hence wage rates; it absorbs savings and capital, which are scarce in poor countries; and in the medium term, it reduces employment for working-age populations, which are growing fast even in HIV/AIDS-affected countries.

MAXIMISING THE HEALTH BENEFITS OF AGROTECHNOLOGY

Pro-poor agrotechnology can offer long-lasting health benefits to the poor. Since the 1980s, however, agrotechnology has become less pro-poor. Research has moved to the private sector without adequate pro-poor changes in the incentive structure facing this sector or in public-private partnerships. This shift has reduced yield and employment growth in smallholder food production, and concomitantly the impact of agrotechnology on poverty reduction and health.

What policies can enable agrotechnology to accelerate its thrust toward sustainable human development, less poverty, and better nutrition and health? Overall, policies should be based on the recognition of the mutual linkages between agricultural research, discovery, and diffusion on the one hand, and health, education, and empowerment on the other. Investing in health will benefit health not only directly, but also indirectly through the adoption of pro-poor, pro-health agrotechnology. Likewise, investing in appropriate agrotechnology will not only stimulate agricultural progress, but will also benefit health and poverty reduction.

In the health sector, rural health services should be improved and action taken to anticipate and reduce any negative health effects of agrotechnology. As for agrotechnology, if it is to improve its promise of enhancing human nutrition, health, and wealth, it is crucial to expand applied and basic agroscience in the international public sector. An important example is research to increase the micronutrient content of the main food staples, such as provitamin A–rich orange-fleshed sweet potatoes (conventionally bred) and golden rice (transgenic), recently expanded in the HarvestPlus programme of the Consultative Group on International Agricultural Research (CGIAR).

In addition, for the health needs of small farmers and labourers—as well as poor consumers—to influence research decisions, governments need to develop:

1. institutions and incentives to promote such people's participation and communication with the formal research community;
2. competition among private research providers; and
3. public research in activities that respond to farmers' needs but are unlikely to attract formal private research.

18

The Agricultural Extension in India

INTRODUCTION

The well-being of the rural population worldwide is invariably linked to the performance of the agricultural sector and to the sector's ability to cope with the challenges that result from rising population pressures, changing demand for food and agricultural products, resource scarcity, climate change, and greater production uncertainty. The World Development Report 2008 (World Bank 2007) emphasises agricultural extension as an important development intervention (1) for increasing the growth potential of the agricultural sector in the light of rising demand-and supply-side pressures; and (2) for promoting sustainable, inclusive, and pro-poor agricultural and hence economic development. The call for agricultural extension services is made at a time when the under-utilisation of the productivity and growth potential of the agricultural sector for development poses a severe threat for achieving food security and for further reducing (rural) poverty.

The degree of institutional and economic development determines the scope and types of agricultural extension services and the ways in which these services are provided and financed (Anderson 2007). Ideally, the design of the service provision and funding arrangements reflects the inverse relationship between the stage of economic development and the importance of extension for agricultural development and poverty reduction. The access to well-defined extension services is more important for economies in which agriculture is a major or declining source of economic growth (agriculture-based and transforming economies) but less important in economies in which agriculture is a minor source of economic growth (urbanised economies).

A prominent example for the group of transforming economies is India. The transition from an agriculture-based to a transforming economy was initiated by macro-economic and non-agricultural reforms in the early 1990s, which triggered unprecedented high growth in non-agricultural (urban) sectors.

At the same time, a weak, ineffective, and inefficient extension system and the consequent deficits in knowledge and technology development and dissemination constrained agricultural-sector growth, which in turn caused the share of agriculture in aggregate income to contract from approximately 31 per cent in 1993 to about 19 per cent in 2003-05. In addition to constraining agricultural sector growth and thus rural development, a weak and ineffective extension system increased the exposure of the agricultural sector to the effects of high population growth, shifts in product demand, natural resource constraints, climate change, and HIV/AIDS, among others (see, for example, Birner et al. 2006; Anderson 2007; World Bank 2007).

In order to meet these challenges, India's extension system has experienced major changes since the late 1990s in governance structures, capacity, organisation and management, and advisory methods.

The changes involve the decentralisation of extension service provision to the local level, the adoption of pluralistic modes of extension service provision and financing, the use of participatory extension approaches, capacity training of farmers to express their demands, and capacity training of service providers to respond to the demands of farmers, among others (Rivera, Qamar, and van Crowder 2001; Birner et al. 2006; Birner and Anderson 2007; Anderson 2007). The reform initiatives reflect the view that improvements in agricultural productivity require demand-driven and farmer-accountable, need-specific, purpose-specific, and target-specific extension services.

Birner et al. (2006) argue that there is no single optimal or best model for providing need-specific, purpose-specific, and target-specific extension services. The ultimate choice of the agricultural extension approach depends on: (1) the policy environment; (2) the capacity of potential service providers; (3) the type of farming systems and the market access of farm households; and (4) the nature of the local communities, including their ability to cooperate. Different agricultural extension approaches can work well for different sets of frame conditions. In order to use extension approaches that best fit a particular situation, the agricultural extension system has to be sufficiently flexible to accommodate the different options. To this end, the recent agricultural-sector reforms have been geared toward creating a demand-driven, broad-based, and holistic agricultural extension system. This has involved the design and introduction of a multitude of integrated measures that, on the demand side, enable service users to voice their needs and hold service providers accountable, and on the supply side, influence the capacity of service providers to respond to the needs of the extension service users (that is, the farmers).

REFORM INITIATIVES IN AGRICULTURE

This Chapter analyses India's major reform initiatives implemented to create a demand-driven, broad-based, and holistic agricultural extension system. The reform projects are studied with regard to: (1) the governance structures of the agricultural extension system and the institutional arrangements for funding and providing agricultural extension services; (2) the actions taken to improve the capacity of extension service providers and users to supply or demand agricultural extension services; and (3) the methods of providing agricultural extension services. In addition, the paper evaluates the effects of the agricultural sector reform initiatives on access to and the quality of agricultural extension services. The ultimate objective is to gain a view on what works, where, and why in improving the effectiveness of India's agricultural extension system, to identify measures that strengthen and improve agricultural extension service provision, and to reveal existing knowledge gaps.

The present Chapter is informed by the analysis of Birner and Anderson (2007) who review India's Policy Framework for Agricultural Extension (India, Ministry of Agriculture 2000) with regard to its strategies for making extension demand driven. Within this framework, Birner and Anderson (2007) identify options for providing and financing extension services, which could help to address market failures and state and community failures, and they discuss the (theoretical) relevance of the strategies to India's agricultural extension policies. In contrast to Birner and Anderson (2007), this paper does not analyse the policy framework under which agricultural reforms are implemented but studies the programmes through which the extension system is being reformed.

Since the objective is to identify what works where and why in strengthening the agricultural extension system, the discussion emphasises agricultural reform initiatives for which performance evaluation or impact assessment studies are available. In addition, the chapter discusses the obstacles to the implementation of the different reform programmes and emphasises methodological limitations in the performance assessments of the reform initiatives.

The remainder of this chapter is structured as follows. Section 2 presents the conceptual framework for the analysis of India's recent set of agricultural extension reforms. Using the conceptual framework in section 2, section 3 analyses the recent agricultural reform initiatives. Major attention is paid to the demand-and supply-side elements of the reform programmes. Section 4 reviews existing performance evaluation and impact assessment studies that identify the implications of the agricultural extension reforms for the quality of and access to agricultural extension services. Section 5 presents obstacles to the implementation of reforms. Section 6 discusses the methodological limitations of existing performance and impact assessment studies and identifies the major knowledge gaps that prevail with respect to: (1) the real effects of the reform initiatives; and (2) what works where and why in promoting agricultural development. Section 7 summarises the main results of the literature review and discusses policy implications.

CONCEPTUAL FRAMEWORK

In order to analyse India's recent agricultural extension reforms, we apply the conceptual framework in Figure 18.1, first presented in Birner and Palaniswamy (2006). The framework identifies the major governance structures, organisational and managerial characteristics, and frame conditions (for example, socio-economic characteristics) by which public-sector extension reforms can improve the organisational and managerial performance of service provision, lead to better public-sector governance outcomes, and generate sustainable pro-poor development. Public-sector governance outcomes can be evaluated in terms of the efficiency, effectiveness, and long-term sustainability of service provision, regulatory quality, rule of law, the degree of corruption, and equity aspects.

Throughout this chapter, the agricultural extension reforms and their impacts are analysed in terms of the underlying supply and demand components of extension service provision and related reform efforts. Demand-side approaches of public-sector service reforms aim at improving the ability of the private sector (such as farm households and profit-oriented firms) and the third sector (such as non-governmental organisations, farmers' organisations, and rural women's groups) to demand better governance and to hold public officials accountable by strengthening the voice of clients. To this end, demand-side approaches include policies that increase information and coordination in voting, strengthen the citizens' right to information, and improve the credibility of political promises. Demand-side approaches of rural service provision also involve policies that promote the political decentralisation of service delivery to local governments, reserve seats in local councils for women, and advocate participatory planning and implementation methods, among others. Figure 18.1 indicates

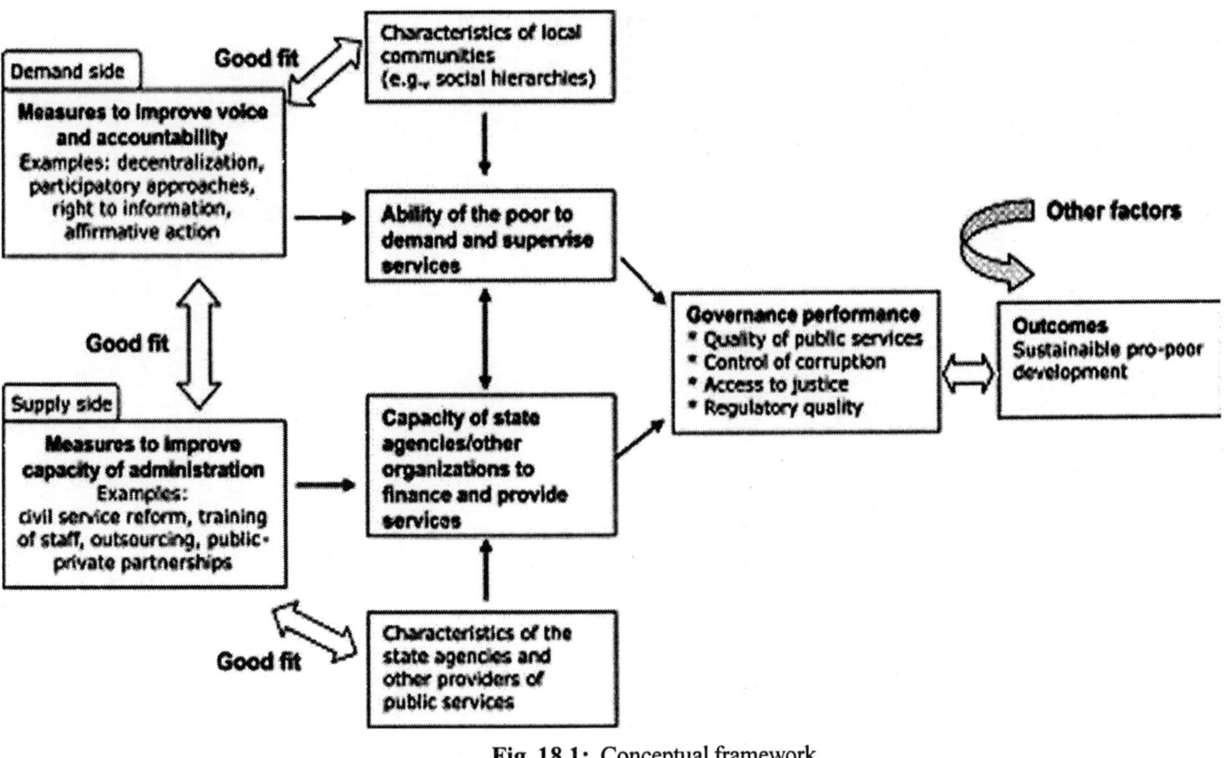

Fig. 18.1: Conceptual framework

that demand-side reforms are likely to be more effective if they directly address socio-economic and socio-cultural obstacles that prevent citizens from exercising their voice and demanding accountability.

Strategies to strengthen the demand side of rural service provision will have little effect if they are not accompanied by strategies to increase the capacity of service providers to finance and deliver the respective services, to apply the rules of law and regulation, and to control corruption. Supply-side approaches to public service delivery reforms include the administrative and fiscal decentralisation of service delivery, public expenditure management reforms, training programmes for public officials, changes in procurement and audit procedures, and efforts to coordinate the activities of government agencies and departments. Another popular supply-side approach reduces the tasks that are performed by public-sector agencies. The respective strategies include outsourcing of service provision to organisations in the private and third sectors, public–private partnerships, pluralistic forms of service delivery, devolution of authority to user groups, and privatisation. Recent reform trends emphasize the need for the state to play a coordinating and facilitating role and to create an enabling environment for the private and third sectors.

Supply-side approaches also include strategies for cost recovery that aim to improve the financial sustainability of service provision and to strengthen the incentive for clients to demand better services.

The usefulness of cost recovery schemes can be debated on equity grounds, especially in absence of specific measures (vouchers for low-income households, for example) that address this concern.

Figure 18.1 suggests that the success of demand-and supply-side reforms depends on the extent to which reform strategies address the sociocultural characteristics of local communities (social hierarchic structures, for example) and the bureaucratic characteristics and incentive structures of public-sector service providers (such as moral and ethical standards and elite capture). Because local communities and service providers differ in terms of characteristics, a one-size-fits-all reform approach is an inadequate mechanism for improving rural service provision for rural development. In addition, the structure and scope of reforms also depends on the feasibility of reform implementation. Rather than engaging in ambitious reform programmes that address all service delivery problems at the same time, it is often necessary to concentrate initially on those reform elements for which political support can be built (Levy 2004). These relationships suggest that reform approaches should center on principles of best fit rather than best practice.

The conceptual framework in Figure 18.1 points to the role of decentralisation as a governance mechanism to improve the quality of and the access to basic services, infrastructure, and legal and regulatory structures. The importance of decentralisation for good or responsive governance is attributable to the positive effect of decentralisation on the efficient use of resources, with efficiency gains arising from: (1) the functional, financial, and administrative autonomy of individual government tiers; (2) role clarity; (3) people's participation; (4) accountability; and (5) transparency, among others (see, for example, von Braun and Grote 2002). At the same time, decentralisation ensures that each government tier performs those tasks in which it has a comparative advantage. The actions taken by the different government levels are then complementary to each other, with the actions being separated by clear boundaries.

DECENTRALISATION AND AGRICULTURE FAILURE

Decentralisation is also subject to weaknesses and is not an accountability mechanism a priori. In fact, economic theory suggests and empirical evidence shows that decentralisation can reduce accountability and increase corruption. Furthermore, decentralisation may lead to welfare losses because of local elite capture and administrative failures (World Bank 2004b). Local elite capture implies that a small share of the population with a disproportionate share of political and economic power resists the changes from decentralisation and participatory policies because of their perceived undue influence on established power relationships. Local elite capture is associated with inefficient use of resources, inefficient and ineffective targeting of public expenditures and transfer programmes, and inefficient delivery of public goods. These aspects amplify inequalities in income distribution, with consequent threats to the economic and political stability of regions. Political and economic tensions may prevail not only within regions but also between them. In fact, they might be stronger if decentralisation reinforces regional asymmetries in income distribution and income growth in the absence of cross-regional income compensation schemes.

19

Agro-Hydrology and Irrigation Efficiency

INTRODUCTION

In irrigation systems where return flows re-enter a fresh water resource and are of reusable quality, water is only saved over the long run through water conservation where the evaporation or evapotranspiration (ET) components are reduced. However, issues of stream flow reduction and time lags can be important.

In cities, the investments in costs for treatment of water and distribution capacity, degradation of groundwater must be considered in addition to when or whether excess water applied returns to a fresh water resource for reuse. Conservation programmes may not save "real" water, but only change the distribution of the resource in space and time. In these cases, the public investment is not well spent. These realities are understood by most hydrology and water resources professionals and by water law professionals working in some states, most notably Colorado and Idaho, principally because the large numbers of water transfers and conflicts in the past have necessitated the recognition of the law of conservation of mass and adherence to hydrologic quantification. However, many politicians, civil servants and special interest groups do not understand the hydrologic context of water utilisation efficiencies and composition.

Some water use terms such as the evaporated fraction, reusable fraction, non-reusable fraction and depleted fraction are discussed that can help the user and the public develop improved visual understanding of the hydrologic context and true impacts of water use and conservation programmes.

Fundamental questions are provided to help quantify the context of the water conservation programme and the impacts from a hydrologic viewpoint.

Water conservation activities are frequently encouraged within irrigation systems and municipalities, especially during periods of drought. The objectives of many irrigation water conservation programmes have been to increase irrigation efficiencies with the expressed purpose of reducing gross diversion requirements. The intent during droughts is that less water will be depleted from a limited resource. However, an incidental effect of increased efficiency is that a larger fraction of the resource is evaporated.

The intent of long-term conservation programmes is that more water will be made available for other users.

However, programmes to reduce diversion requirements must have both a regional and local interpretation from a hydrologic and conservation of mass viewpoint. Water management principles used to guide society's water use objectives require terms and definitions that clearly describe the effects of various water uses, both consumptive and non-consumptive, within a hydrologic system.

Some water use terms such as the evaporated fraction, reusable fraction, non-reusable fraction and consumed fraction are discussed in this paper. These terms are useful to both users and public in developing improved, rational and visual understandings of the hydrologic nature and impacts of water use and conservation programmes. In situations where the non-evaporated components of irrigation diversions return to the fresh water resource for reuse by others, conservation programmes may not stretch water supplies or "save" water in the region, especially in the long-term, and especially where the initial source is from ground-water. In some instances, where water is abstracted from streams, irrigation water conservation programmes can actually be "ET sustainment" programmes, since they may sustain a more "consumable" water supply for one city or project at the potential expense of downstream projects, cities and perhaps the environment. Water conservation programmes should fundamentally be evaluated in the context that, in general, the only real loss of water from an irrigation project is by the process of evaporation from open water surfaces, evaporation from moist soil, transpiration from vegetation, and flows into saline sinks. Fundamental hydrologic concepts and questions are described that can help planners and managers to establish the context and impact of individual conservation programmes in the near and long term. Case studies in the United States are given.

FUNDAMENTAL PRECEPTS

There are fundamental precepts that govern the ability to conserve water and the ability to create "new" water by a conservation programme. These are:

- The law of Conservation of Mass. The Law of Conservation of Mass suggests that matter can not be created nor destroyed. In the context of liquid water, the law suggests that liquid water, while remaining as liquid water (and not evaporated) can not be created nor destroyed. Thus all nonevaporated components must be "somewhere" and must reappear "somewhere" (figure 19.1).
- The reality that 99 per cent of the earth's landmass is underlain by ground-water (Freeze and Cherry, 1979) impacts the "loss" of water. All deep percolation "losses" are not "lost" to the hydrologic system, but seep downward vertically to the groundwater. After entering the saturated groundwater system, the liquid moves with the groundwater laterally, at some velocity determined by hydraulic gradient and geology, until it discharges to a surface water source.

Irrigation Hydrology

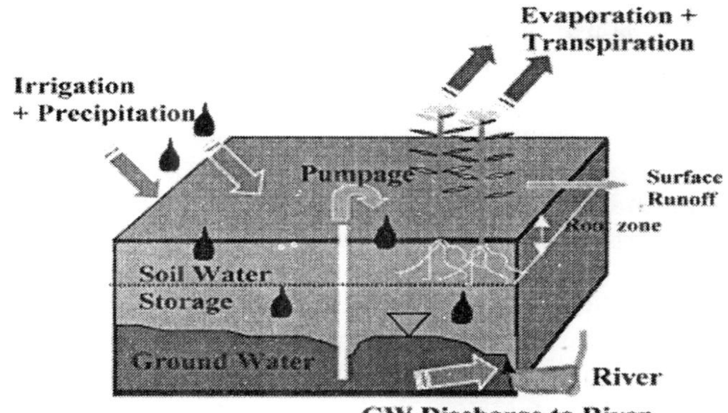

Fig. 19.1: Schmatic showing hydrologic partitioning of water and flow paths between components

Thus, the only way to really create "new" water is to reduce the water that is degraded to the point where it is not usable by anyone else downstream or to reduce the evaporated component of the diversion (i.e., reduce the evapotranspiration, ET).

Basic Hydrology and Law of Conservation of Mass

There are saturated ground-water bodies lying beneath the earth's surface almost everywhere in the world.

These ground-water bodies have had thousands of years to develop, and have built up to an equilibrium point so that ground water flows freely by gravity, if it is unconfined, to a lake or stream system (or to the ocean, if nearby) where it discharges. Unless they have been overdrafted by pumping, most ground-water systems are in equilibrium with surface water systems. Most streams exist during periods of low surface runoff because a ground-water table feeds the stream. The addition of water to a ground-water system is, over the long term (perhaps tens of years or less), balanced by similar amounts of outflow to a surface system. The flow process is controlled by gravity, is automatic, and is inevitable, i.e. part of the basic hydrologic equilibrium.

The primary consumption of water within an irrigation system is by the process of evaporation from open water surfaces, evaporation from moist soil and transpiration from vegetation. The combination of this evaporation and transpiration is termed ET. The transpiration of water is a necessary and inevitable part of vegetative growth and agricultural production (Seckler et al., 2002). In addition to ET, water that is returned to a saline water body or that is severely degraded in quality is essentially lost as a freshwater resource. All other water diverted by an irrigation system remains in liquid form and will ultimately return to a freshwater system. The return of diverted water to the system is a natural, diffusive process that is nearly impossible to control, because remaining liquid water must obey the law of gravity and the law of conservation of mass. Gravity brings non-evaporated water back to a stream, ocean or aquifer system.

A consequence of reducing water diversions is almost always a reduction in return flow back to the resource. Therefore, the quantity of net consumption by an irrigation system may be largely unchanged by a conservation programme. To effectively create "new" water in a regional context, unless directly upstream of a salt sink, a conservation programme must in some way reduce evaporation or ET or improve return flow quality, and not simply reduce diversions. Reductions in the direct consumption of water are usually in the form of reducing areas of phreatophytes or wetlands along canals, collection ditches, or in areas of shallow, ground-water seepage to the soil surface. Wetlands and phreatophytes created by irrigation are often considered to be of value for wildlife habitat and may be lost when water conservation practices are implemented. Reduction of crop ET will almost always reduce turf quality or crop yields, unless evaporation from soil is reduced without reducing plant transpiration.

Water conservation efforts in irrigated agriculture frequently focus on:

- Improving the infrastructure of the irrigation system so that water can be controlled more easily.

- Improving the management of the irrigation system so that the right amount of water is applied at the right time.

It is important that irrigation improvement procedures be evaluated to show when and how water is actually saved by the conservation programme. Guidelines for preparation of conservation plans must include procedures for describing hydrologic components and interactions within and beyond irrigation system boundaries, with descriptions and examples of how to assess whether evaporation or ET can be reduced within the system or "return" flows into saline systems can be reduced, thereby achieving real conservation of water and the creation of an enhanced water supply. Unfortunately, it is common to draw "lines" around system boundaries and to neglect the real interconnections between in-system "losses" and existing river system gains.

Burt et al. (1997) suggested a systematic way to examine the water that is diverted for irrigation. In evaluating the performance of irrigation systems, one has to be careful to establish appropriate physical boundaries and time frames, since water is often in transit or in temporary storage. One can only evaluate the performance of an irrigated area by examining the irrigation water when it leaves the defined boundaries of interest. The applied irrigation water can be placed into several categories:

1. Water consumed by the crop within the area for beneficial purposes.
2. Water consumed within the area under consideration but not beneficially.
3. Water that leaves the boundaries of the area under consideration but is recovered and reused by the same party or by a "downstream" party.
4. Water that leaves the boundaries of the area under consideration but is either not recovered or not reusable.
5. Water that is in storage within the boundaries.

The Burt et al. categorisation can help to identify improvements that may truly conserve water. The following paragraphs explain and discuss the five categories above (Clemmens and Allen, 2005):

1. *Category 1* water is used to produce a crop and is consumed by transpiration from stomates of leaves or from soil beneath or between plants. Reducing this water consumption generally reduces yields or marketable products. Evaporation from soil averages from 10 to 20% of total consumption (ET). Buried drip irrigation claims a benefit here, since the soil surface is sometimes not wetted and thus the evaporation component can be reduced with only slight increase in transpiration. However, claims of large savings in water consumption are physically unreasonable, since buried drip systems can generally only reduce evaporation by about one-half and consequently can create savings of only 5 to 10 per cent of total consumptions and likely only 2 to 5 per cent of total water diverted.

   One option is to fallow land or take it out of production. This results in real water conservation, particularly in arid environments, since ET is substantially reduced. However, fallow land can have negative environmental consequences if not done properly (soil salinisation, wind erosion, etc.). Genetic improvements to produce more crop with less water are possible in some cases, but usually more crop is produced with the same water (or the same yield on less acreage and thus less water). Outside of genetic improvements, reductions in water consumption are limited because crop biomass accumulation is through photosynthesis where carbon dioxide enters the plant through the same stomates as water vapour exits. Therefore, the ratio of yield to transpiration in a specific climate is limited by the physiology of the crop variety (Seckler et al., 2002).

2. *Category 2* includes water that is lost through evaporation from soil away from crops and open water surfaces and evapotranspiration from weeds, trees, phreatophytes, etc..

Many of these losses are difficult or costly to avoid, but there are opportunities in some cases. Some of the lost water may have environmental benefits through providing water to sustain trees, windbreaks, incidental wetlands etc. that are attractive to wildlife.

3. *Cateorgy 3* water, which is recovered and reused, should not be a target for water conservation.

   The water is already being reused, so that reducing this from entering the irrigation system often has no real benefit. There are exceptions, for example where it reduces supply costs or improves the quality of the water, or where leaving the water in a reach of a stream has an environmental benefit. In some cases, the delay in unconsumed irrigation water returning to the stream has an environmental benefit by buffering streamflow during periods of drought or winter.

4. *Category 4* water is the primary focus for water conservation efforts. Here water is unrecoverable because of large depths to groundwater or is not reuseable because of salinisation or entry into a saline body such as the ocean or brackish aquifer. A reduction in water applied for irrigation means that it can be used for other purposes. It is truly water saved. However, there are cases where this water provides environmental benefits, such as mitigation of salt water intrusion underground, even though it has no human use.

5. *Category 5* is irrigation water that is in storage within the area of consideration and not yet considered to be used. The ultimate fate of this water has yet to be determined. The delay caused by storage may have positive or negative economic and hydrologic impacts.

Appropriate Reasons to Conserve Water by Increasing "Efficiency" (Uniformity) of Water Application—

The following are appropriate reasons why irrigation projects or systems or cities should conserve water:

- Reduce costs for treating water;
- Reduce costs for pumping water;
- Reduce costs for added distribution capacity in an area of growing population or demand;
- Reduce leaching of fertilisers and other chemicals and degradation of ground-water;
- Sustain flows in specific segments of streams that are threatened by low flows or thermal increases and where "nonevaporated" components of diverted water bypass the stream or return at a less valuable time;
- Where "non-evaporated" components of diverted water flow into a saline system (ocean, saline lake, or brackish groundwater) and are therefore non-recoverable or contaminate streams downgradient;
- Where water is abstracted from a deep, confined aquifer, but the "non-evaporated" components of abstracted water percolate to a more shallow unconfined aquifer, thus changing the distribution of water between the aquifers in an undesireable way.

In agriculture:

- Reduce waterlogging and improve salinity control;
- Enhance equity among users;
- Maximize the total fraction of water delivered to crops to increase crop yields;
- Reduce soil erosion.

Inappropriate Reasons to Conserve Water by Increasing "Efficiency" (Uniformity) of Water Application.

The following are inappropriate reasons to initiate a water conservation programme:

- To create "new" water downstream in regions where return flows already re-enter the water resource at an appropriate time;

- To enhance streamflows for long distances downstream where return flows already re-enter the water resource at an appropriate time;
- To extend the life of an unconfined aquifer where return flows already re-enter the aquifer with acceptable quality.

BENEFITS OF LOW EFFICIENCIES

The term irrigation efficiency (IE) has traditionally been defined as the ratio of the sum of beneficial consumption and leaching to gross diversions (Jensen, 1967; Bos, 1985). The following are benefits of low efficiencies of water application (i.e., overirrigation or poor distribution uniformities) that are realised in a number of situations (Allen et al., 1997). These benefits are generally incidental and fortuitous and may not be intentionally designed into a water management programme:

- Recharge to unconfined aquifers;
- Dampening of flood flows or redistribution of flows over time (due to reentry of return flows with some time lag and dampening);
- Augmentation of streamflows during droughts (due to re-entry of return flows created by diversions during periods of higher flow). The augmentation of diffusive return flows by groundwater may help cool streamflow and benefit biota;
- Incidental ground-water recharge near oceans may help reduce salt-water intrusion;
- Creation of wetlands.

The appropriate reasons for conservation programmes listed above are all valid reasons and goals for water conservation efforts, but they should be worth the price paid to obtain them. Many improvements may not conserve water on a regional basis, since ET of irrigated fields, lawns or golf courses is normally not reduced in these types of "conservation" programmes. In fact, ET may actually be increased due to improved uniformity and more careful control of water application. Therefore, water conservation efforts on the local scale may ultimately increase water consumption both on a local and regional scale.

There are hydrologic systems where nearly one hundred per cent of the water is being productively consumed due to natural reuse within the system. Total consumption in such cases cannot be increased past 100 per cent, nor can altered practices designed to "increase irrigation efficiency" in such a system yield additional water to be used by new diverters without reducing the consumption (i.e., evaporation) of current users. Use of the term "irrigation efficiency" has caused a dichotomy between the physical situation of the hydrologic system and the public's and government's perception of the physical nature of water management. These incorrect views are pervasive and strongly held. Billions of dollars have been proposed for investment to correct for low irrigation efficiencies with the intent that water problems will be solved. The public has been convinced that selected investments and penalties imposed on irrigation will free up vast amounts of water for other uses. Only a fully rational approach to water management can minimise the conflicts that arise between municipal, industrial, environmental, recreational, aesthetic, and agricultural uses of the finite fresh water supply.

DEFINITION OF WATER CONSUMPTION TERMS

An improved, graphic image of the hydrologic and basin-wide effects of irrigation is possible when the disposition of water within an irrigation project is described in terms of "fractions." Definitions based on fractions have been proposed by Jensen (1993), Willardson et al. (1994), Allen et al. (1996, 1997) and Molden (1997) and Molden and Sakthivadivel (1999) for assessing the impacts of fresh water diversions by users of water resources, including irrigated agriculture, municipalities, industry, and ecological interests.

The new terms are intended to encapsulate clearly the impact of any and all types of water use on actual physical losses of utilisable water from the affected hydrologic system. Unlike most efficiency terms, the proposed methodology and terms: (a) are appropriate for evaluating water allocation, water use, and related management options; (b) are consistent and appropriate for all water uses, not only for irrigation and a narrow evaluation of irrigation practices; and (c) can be clearly understood conceptually and in terms that can be correctly applied by people engaged in the water allocation/use/ management debate.

Application of such terms will help to clarify what the allocation of water to various uses at various locations in a hydrologic system actually means in terms of the total water supply.

A change from using "efficiencies" to using "fractions" to describe water use eliminates many misunderstandings. Fractions are used in many applications to describe what proportion of some quantity has been applied to a particular use. Use of a fraction evaluation instead of an "efficiency" prevents the occurrence of a serious logic error in describing or evaluating the management of water. Jensen (1993) discussed the need for a change in the ways that water use is described, and has also advocated moving away from use of the term efficiency in irrigation.

Figure 19.2 shows a matrix of uses and disposition of irrigation diversions categorised as beneficial and non-beneficial and as consumptive and nonconsumptive as described by Clemmens et al. (1995), with enhancements to the water disposition categories by Allen et al., (1996, 1997). The figure illustrates relationships among the following fractions proposed to describe the hydrologic disposition of irrigation diversions. The fraction terms are defined as follows:

EVAPORATED FRACTION

The evaporated fraction (EF) is the fraction of an irrigation diversion that is consumed through evaporation or evapotranspiration:

$$EF = \frac{Q_{ET}}{Q_{Div}} \quad \text{------------------------------} 1$$

where QET = quantity of diversion consumptively evaporated (or transpired) by the water use process (for example, irrigation) and QDiv is the total diversion of water

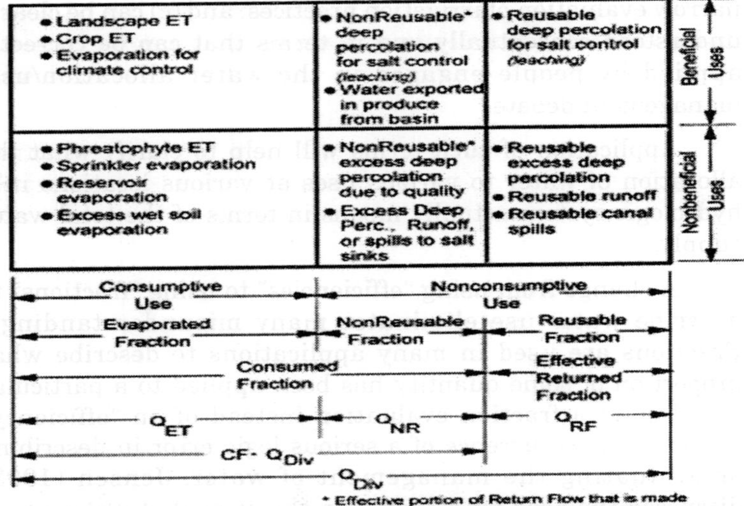

Fig. 19.2: Use categories (consumptive and non-consumptive and beneficial and non-beneficial) and fractions describing the disposition of irrigation diversions (after Burt et al., 1977 and Allen et al., 1996, 1997).

to the specific process. Besides ET from landscapes or cropped fields, QET includes evaporation from ponds, canals, reservoirs and seeps, and water evaporated from riparian vegetation and wetlands created by irrigation return flow or seepage. EF is similar to the irrigation consumptive use coefficient term introduced by Jensen (1993), except that EF may also include evaporation external to the primary process.

NON-REUSABLE FRACTION

The non-reusable fraction (NRF) is defined as the fraction of a diversion that is not evaporated, but is no longer available for reuse by other water users due to entry into a saline system (ocean, brackish water bodies, or saline aquifers) or due to degradation in quality to the point that it is economically non-resuable, or is physically beyond economic recovery:

$$NRF = \frac{Q_{NR}}{Q_{Div}} \text{ -2}$$

where QNR = quantity of diverted water that is still in liquid form, but has been made non-recoverable due to the physical manipulation by the user (diverter). The NRF represents the fraction of QDiv that could conceivably be made available to other users, in addition to reductions in non-beneficial ET, through conservation efforts, without reducing crop yields. Non-recoverable water that results from a particular use should be identified and charged to that use.

CONSUMED FRACTION

The consumed fraction is defined as the fraction of total diversions that are consumed, i.e., no longer available to any other user during any future time period. The consumed fraction includes the evaporated fraction and the non-reusable fraction, since physically, these two fractions are "consumed" in the context of the fresh water resource. The consumed fraction (CF) includes any water exported from the basin:

$$CF = \frac{Q_{ET} + Q_{NR} + Q_{exp}}{Q_{Div}} \approx EF + NRF \text{------------} 3$$

where Qexp = water that is exported to outside the hydrologic basin. An example is water contained in fresh fruit that is transported from a basin, or in the case of production of bottled water or other beverages, the water contained in the beverage, assuming that the beverage is not consumed within the basin.

In the definition of consumed fraction, the term "consumed" means that the CF fraction of the diversion is truly consumed or otherwise transformed so that it is no longer reuseable by any other future user within the basin. The consumed fraction of diversions either undergoes a phase change (evaporation), is exported outside the basin, or enters a non-reusable state due to extreme salinisation pollution, or uneconomically recoverable location, any of which make the water non-reuseable by anyone else. It is important that the reader realise fully that water diverted by an irrigation project or any other user is not "consumed" unless one of the transformations occurs (transformation from liquid to vapour or entry into a non-reuseable quality state). The user should be considered responsible for the quantity of the water resource which, on a basin-wide scale, is the product CF • QDiv.

REUSABLE FRACTION

The reusable fraction (RF) represents the fraction of the diverted water that returns to the water resource for subsequent reuse by others:

$$RF = \frac{Q_{RF}}{Q_{Div}} \qquad (4)$$

where QRF is the quantity of diverted water that is reusable by other users. QRF naturally reenters the fresh water system.

Importance of Local Hydrology and Location within a River Basin, some irrigation projects are located close to the ocean or directly upstream from other saline water systems such as saline lakes or saline ground water sinks. In these situations "return" flows from irrigation projects enter these saline systems and are truly lost for additional consumption by humans. In these situations, reducing diversions by enacting water conservation programmes may allow upstream users to divert and consume more water, thereby increasing the total beneficial consumption of the water resource.

In areas where excess diverted water percolates through soil profiles and picks up salt from soil and rock, return flows from deep percolation increase the total salt load of the receiving water resource and may reduce its economic usefulness. In these cases, reducing diversions and return flows by increasing irrigation uniformity and reducing excessive applications may increase the effective water supply.

Willardson and Allen (1998) delineated upper, middle and lower regions of river basins and discussed impacts of water conservation in each location (figure 19.3) to assist in assessing impacts of low "efficiencies" (i.e, consumed fractions) on downstream and other users. In general, the need for conservation programmes and impact of such programmes increases as one moves downstream (toward the ocean). In low regions of basins, the NRF (non-recoverable fraction) of diverted water generally increases due to proximity to saline systems.

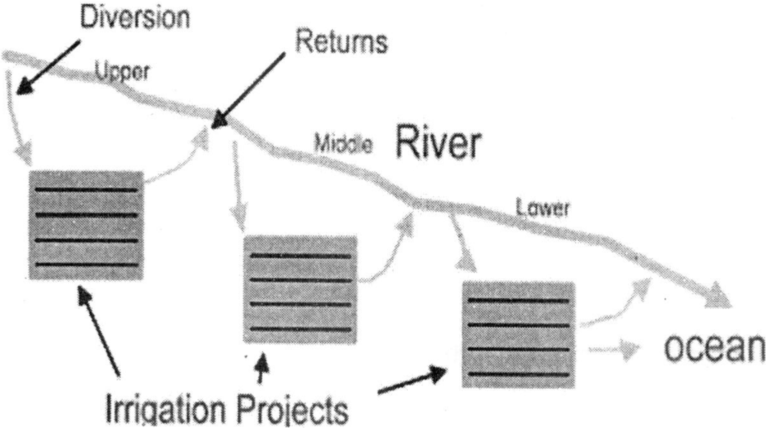

Fig. 19.3: Categorisation of a river basin into upper, middle and lower regions for assessment of impacts of water use, consumption and return flow (from Willardson and Allen, 1998)

Non-reusable quantities of water arise in other uses of the water resource besides irrigation. For example, water allocated to "wild rivers" in northern California is in general not recovered for other uses and runs directly into the ocean and becomes non-recoverable. Such water has a very low evaporated fraction (EF) but has a very large non-reusable fraction (NRF) and consequently, a large CF. Such uses of water should be described in the same terms as for irrigation so that the public understands the impact on total available fresh water in terms consistent with the descriptions used for other uses. A low CF for a city high in a watershed permits a large fraction of the returning water to be reused downstream after natural bioremediation and/or water treatment. A high CF for a large coastal city will result if nearly all of the sewage effluent (reusable under some circumstances) becomes non-reusable when it is discharged directly into the ocean. It does have high benefit, however, if injected to groundwater to reduce seawater intrustion. In some situations, it is "good" to have a "low" CF, since this means that much of the diverted water returns as a fresh water resource for subsequent reuse. However, because a low CF is equivalent to a low "efficiency", the latter term typically gives a falsely negative impression to the public.

EFFECT OF SCALE

Fractions can be calculated for any scale of interest. In the case of irrigation, this is typically at the field, sub-project, project or basin scale. Generally, the non-reusable or QNR quantities of water for fields or conveyance systems in upper regions of irrigation projects are small relative to QDiv and QEP, especially if hydrology and elevation promote convenient and timely reuse of water or return of water to the stream or to a recoverable ground-water system.

An example of this is the Little Willow Irrigation District in southwest Idaho where the geology and topography of the long, narrow mountain valley containing the irrigation project promotes rapid reentry and reuse of surface and subsurface return flows within the project boundaries.

Irrigation "efficiencies" (or more correctly, consumed fractions) of individual farms average only 0.30, but the total project irrigation "efficiency" or consumed fraction for the project is 0.60 due to the reuse of water. The remaining 40 per cent of diversions not consumed by the Little Willow project (i.e., 1 - CF) return to the surface water resource below the irrigation project and are diverted by other downstream water users, making the RF for the basin very high.

The acceptable magnitude for NRF for an individual field, lawn, golf course or other use may be different from the system-wide average NRF. Actual NRF may be low for fields or conveyance systems in upper regions of irrigation projects where the opportunity for reentry and reuse of deep percolation, surface runoff, spills and seepage is high. NRF may be high for similarly irrigated fields or conveyance systems near the lower portions of irrigation projects when percolation or runoff directly enters the ocean or brackish water bodies.

FUNDAMENTAL QUESTIONS

There are fundamental questions that one should ask when evaluating the potential impacts of a "water conservation programme" on ultimate water savings and impact. These questions are posed from a hydrologic perspective and adherence to the law of conservation of mass.

1. Where does the delivered water come from? (i.e., is it from a stream, ground-water, or lake?) Where is the location of the abstraction?
2. At what time of the year are the abstractions made? (i.e., what does the abstraction "hydrograph" look like?)
3. Where does the non-evaporated component of any applied water go? At what times? (i.e., hydrograph of flows of non-evaporated components)
4. Where does the non-evaporated water reappear as part of a ground-water or surface water system? At what times? In what quantities? With what quality?
5. What happens in the mean time (between the abstraction and the return to the resource)? What are the consequences of this time lag or spatial lag? (i.e., is there local stream dewatering? Are there junior appropriators without water?)

REASONS FOR ACTION

1. If there are local instream flow needs that are not being met, then reduce diversions with conservation. However, the conservation programme will not create new water for other users outside of the specific system or enterprise. In fact, the conservation programme may be an "ET sustenance" programme at the expense of downstream users and may reduce downstream flows.
2. If the water use near a saline system (ocean, brackish sink, etc) so that non-evaporated components are impaired or lost via quality change, then a conservation programme will have a good hydrologic impact.

3. If there are system capacity constraints or if there is large invested treatment (culinary) or energy costs involved, then a conservation programme should be considered for local economic reasons and may not result in savings to the water resource.

BASIC CONSERVATION QUESTIONS

1. How much of the water abstraction gets consumed or moves beyond local control? (What is the CF?)
2. Who benefits from "wasted" water when it reappears and is recovered?
3. Are current "downstream" users better off by any higher efficiencies created in systems upgradient by a water conservation programme? Are the downstream users benefited quality-wise?
4. Is other water available by other means?
5. Will conservation make "new" water available to other local or within-system consumptive processes so that the net effect of the conservation is even less water downstream? (this is in the opposite direction intended or purported by many conservation programmes, but may be the hydrologic reality).

CONCLUSIONS AND RECOMMENDATIONS

Irrigation is no longer an endeavour isolated from other users of the fresh water resource. For regional water management, determination of the consumed fraction (CF) and reusable fraction (RF) is much more relevant than irrigation efficiency, and the use of these fractions may help to eliminate misunderstandings.

CF is the fraction of diversions that is consumed via evapotranspiration or flow of water to a salt sink. RF is the fraction of diversions that returns to the fresh water resource and can be reused. Emphasizing or promoting conservation programme components that increase efficiencies, without strong caution and guidance concerning whether the RF is already being utilised downstream or from ground-water, may harm both users and the economy. Irrigation enterprises contemplating conservation investments must know whether environmental, economic or landscape health and crop yield benefits stemming from a local conservation programme are worth the cost. The public and other groups that are interested in freeing up water supplies for new uses must know whether a conservation programme will ultimately create new water.

The impact of a given use on the water resource should be expressed in terms of: (a) the fraction of water the use directly consumes; (b) the fraction, by virtue of that use, that is rendered unavailable to other users; and (c) the fraction that is returned to the hydrologic system for reuse. It is understood that the hydrology of irrigation projects and their impact on basin-scale hydrology can be complex due to the wide ranges and variations in geology, mineralogy and timing of ground-water flow systems. Therefore, the quantification of non-recoverable fraction and returned fraction may be difficult in some situations.

However, the use of simple fractions serves as a good starting point for assembling a clear understanding and definition of the hydrologic destiny of fresh water diversions.

Conservation programmes should attempt to reduce the consumed quantity of water, which mathematically is the product (CF • QDiv) where QDiv is the diverted quantity. This reduction requires either reducing evapotranspiration (and thereby potentially reducing crop yields) or reducing any non-recoverable quantities. In reality, many conservation programmes target increasing the "irrigation efficiency" (IE), which may be counterproductive, since, as shown in Figure 19.2, IE • QDiv contains different terms than are present in CF • QDiv. As generally defined (Clemmens et al., 1995), IE • QDiv includes some QRF and omits some QNR.

20

Irrigation Systems and Water Management Programme

INTRODUCTION

At present the low water availability is one of the main constraints in increasing per acre yield and bringing more area under cultivation. The per capita availability of water has reduced from 5650 m^3 in 1951 to 1200 m^3 in 2003 against the international standard of 1500 m^3. It will decrease to 800 m^3 in 2025. Moreover, competition for water is also increasing with the rise in population, urbanisation and industrialisation. The domestic and industrial water uses will increase to 15 per cent by 2025 of the available water resources against 3 per cent in 2000.

The low water productivity is due to low water use efficiency as a result of conventional irrigation methods and poor agronomic practices. Under this situation there is need to utilize the available water more efficiently and wisely alongwith introducing innovative and high efficiency irrigation systems. Therefore, Irrigation Systems and Water Management Programme of Water Resources Research Institute (WRRI) has been devised with following objectives.

OBJECTIVES

- Design, layout and install irrigation systems to improve water use efficiency.

- Adopt high efficiency irrigation systems using locally available materials and techniques.
- Efficient Utilisation of low quality waters.

RESEARCH ACHIEVEMENTS

SPRINKLER IRRIGATION

Sprinkler irrigation systems may be used for all types of soils, topographic conditions and crops. These are labour, water and fertiliser efficient. Sprinkler irrigation systems are most desirable for soils with a high permeability and/or low water holding capacity. However, sprinkler irrigation systems require high pressure pumps. Water Resources Research Institute (WRRI) in collaboration with local pump industry developed multistage pumps which deliver discharge from 2 to 12 lps and produce pressure upto 60 m. Different models of Chinese' rainguns were tested and two models (PY1-30 and PY1-50) have been indigenised in collaboration with local industries. Different sizes of replaceable nozzles have also been developed to cover a range of pressure and discharge under different field conditions. Diesel engine and PTO driven (tractor operated) raingun sprinkler irrigation systems were also developed. In a portable system, a high pressure pump is coupled with a prime mover (electric motor, diesel engine or a tractor) and is mounted on a two wheel trolley which can be moved from one location to another. In semi-portable, the pumping unit is fixed near a water source. The main delivery pipe may be fixed and laterals alongwith raingun are moved from field to field. This type of system is good for small farmers having an area of 2.5 ha.

Models of Raingun (PY1-20, PY1-30, PY1-40 and PY1-50)

Portable Raingun sprinkler irrigation systems

TRICKLE IRRIGATION

Trickle irrigation has great potential especially in water scarce areas and soils with undulating topography and poor water holding capacities. High initial cost of imported trickle irrigation equipment is considered to be the main limitation for large-scale adoption in Pakistan. So WRRI in collaboration with local industry developed low density polyethylene (LDPE) pipes (13 mm to 75 mm diameters) and their fittings. Various types of emitters; spiral, turbo and micro-tubing were tested. The micro tubing emitters performed better under local field conditions due to their easy installation, operation and maintenance. Application uniformity and pressure variation in locally developed trickle irrigation systems were evaluated. The discharge of the micro-tube emitter varied from 7 to 12 l h-1 under pressure of 15 to 19 m. The water application uniformity was above 85 per cent showing satisfactory performance of the system.

SUPPLEMENTAL IRRIGATION

A considerable area of the country (44% of 79.8 million hectares) is outside the Indus basin where rainfall is the only source of water for crops. In the rainfed area, stochastic behaviour of rainfall limits profitable crop production. The yields of major crops are 30 to 50 per cent lower than the national average crop yields mainly because of lack of available soil moisture at the critical crop growth stages. The sowing of wheat crop is often delayed due to uncertain and scanty rains. Furthermore, water for agriculture is becoming scarce. It is likely that it will increasingly be used as supplemental deficit irrigation rather than full irrigation. In rainfed areas where small quantity of water is available in the form of nullahs, ponds, reservoirs or wells, the sowing of crops at right times can be managed by applying small quantity of water through raingun sprinkler irrigation to moist the required depth of rootzone for seed germination. Similarly when the soil moisture depletion has reached more than 80 per cent, the life saving irrigation can be applied with raingun sprinkler irrigation system. The benefits of supplemental irrigation with raingun sprinkler irrigation system were evaluated. The wheat grain yield increased 1.4 to 1.7 ton ha-1 and straw 1.3 to 2.9 ton ha-1 with the use of supplemental irrigation. The highest net benefit was achieved with application of 25 mm irrigation with sprinkler system at 75 per cent depletion of available water. The wheat crop yield increased 45 to 100 per cent over rainfed with 10 to 25 mm supplemental irrigation.

PRESSURE DISCHARGE RELATIONSHIP

The information about pressure and discharge is pre-requisite for designing sprinkler and trickle irrigation systems. The raingun PY1-30 is mostly recommended for irrigation of crops due to its less pressure and energy requirements. The pressure and discharge relationship for PY1-30 have been developed. The system was operated with 6, 8, 10 and 12 mm diameter nozzles and 15 to 55 m pressure. The discharge of raingun increased with increase in pressure. The relationship was developed which is being used for designing of system. The use of PY1-30 raingun with 10-12 mm nozzles and working pressure of 34-40 m an area of 0.21 to 0.26 ha can be irrigated at one setting with application rate of 2.94 to 3.86 mm ha-1. The area and application rate can be increased with PY1-50 raingun. At working pressure of 40 to 50 m, the area coverage is 0.43 to 0.50 ha with application rate of 4.11 to 5.03 mm ha-1 when nozzle diameter varied from 16 to 18 mm.

For trickle irrigation system, locally developed micro-tubing emitters are performing better. So these emitters were tested under pressure heads of 3 to 21 m for different lengths and diameters. The pressure-discharge relationship for micro-tubing emitters has been developed. The relationships are being used for designing of trickle irrigation system.

SURFACE IRRIGATION

Considering importance of the problem, research studies were initiated at Mardan to compare basin and bed-furrow irrigation systems for maize and wheat crops. The average irrigation depth under bed-furrow was about 30 per cent less than basin for maize and wheat whereas, maize grain yield was 42 per cent higher under bed-furrow than basin. Furthermore, the water use efficiency was 67 per cent and 17 per cent higher under bed-furrow for maize and wheat crops, respectively.

Border irrigation system for various stream sizes were tested at five locations under various agro-ecological conditions. Three unit stream sizes of 1, 2 and 3 lps, and three border lengths of 30, 60 and 90 meter were evaluated. The application efficiency of 90 per cent was achieved with stream size of 3 lps m-1 and laser leveled border of 60 m long. The increase in stream size increased the application efficiency mainly due to reduction in advance time and uniform leaching.

SKIMMING WELLS

A thin layer of fresh water lies over saline water mostly in the centre of "DOABS". This fresh water needs to be skimmed without mixing of underlying saline water due to up-coning. As the skimmed water is in small quantities and can not be used for surface irrigation. A collaborative research project was undertaken in SCARP-II, where skimmed water was used through sprinkler and trickle irrigation systems. The four sprinkler and two trickle irrigation systems were installed for sugarcane, wheat, fodder and orchards (Guava and Citrus).

USE OF LOW QUALITY WATER

The farmers are in the process of increasing cropping intensity by using groundwater to an extent of around 50 per cent of total water use in agriculture. In the Indus basin the groundwater aquifers in around 60 per cent areas are of marginal to hazardous quality. A research study was conducted in the SCARP- II area to evaluate the changes occurred in the quality of groundwater with time. It was found that the quality of 12 per cent project area changed from very good (<500 ppm; TDS) quality to marginal quality (>1000 ppm; TDS) in 30 years (1965-1997) indicating deterioration of groundwater quality. The low quality water can be used for salt tolerant crops, trees and shrubs. A research study was undertaken in Haryana, in which irrigation was provided with marginal quality groundwater (EC = 1.36 dS m-1, SAR = 9.2 and RSC = 2.1 meq l-1) to establish fruit trees. In another study, the barren and abandoned land was brought under cultivation using low quality water with integrated water management techniques, such as use of farm yard manure, green manuring, use of gypsum, and sulphur. The janter and kallar grass were grown along with the Guava, Falsa and Beer. A technology and knowledge has been developed for use of low quality water with minimum disturbance of soil health.

WATER MEASUREMENT

Accurate measurement of irrigation water permits more intelligent use of this scarce natural resource. Different types of water measurement devices have been tested and evaluated for water measurement under different field conditions. A "V" notch is the standard water measuring device which is used where surface gradient is available. The "V" notch was used to calibrate cutthroat flume and broad-crested weir. The cutthroat flume can be used under free flow conditions. The broad-crested weir should be used under submerged conditions. The "H" type flumes can handle large and variable flows.

IRRIGATION SUPPORT SERVICES

Due to water scarcity, the awareness for improved irrigation methods among farming community is increasing. The WRRI has initiated an irrigation support services for developmental agencies, NGOs and farmers. The irrigation systems are designed considering soil, water, crop, climate and farmers preferences. On an average about 5-6 farmers visit WRRI monthly and discuss their irrigation management issues.

21

Agro-Technology and Soil Property

INTRODUCTION

Soil-forming processes are very complex and our understanding of this complexity is imperfect. An effective way to treat this uncertainty is to model soil properties as realisations of a random function (Webster, 2000). Geostatistics generates models of these random functions, which are then used to describe variation in soil properties at different spatial resolutions and can be used to predict them at unsampled locations. This approach has been successfully applied to a wide range of soil properties, including soil metals (Goovaerts and Webster, 1994) and the composition of soil microbial communities (Brockman and Murray, 1997; Franklin and Mills, 2003) and various other soil properties (Grunwald, 2006). Predictions from geostatistical models can often form the underlying data used in precision farming (Sadler et al., 1998) or in environmental fate and process models (Corwin et al., 1997). The geostatistical model of spatial variation also underlies linear mixed models for spatial data (e.g. Lark and Cullis, 2004).

The principal underlying assumption of geostatistics is that the stochastic process is stationary. Consider a set of N sampling locations along a line (transect), X=[x1,..., xN], then the data obtained at these locations, z(x1),..., z(xN), are samples from the marginal distributions, D1,..., DN, which are projections of the N-variate distribution D of the random function Z(x). We have only one sample from each marginal distribution DN, resulting in one vector z(X)=z(x1),..., z(xN), drawn from D, which is insufficient to estimate the parameters of D. In geostatistics we make inferences about the random function Z(x) by invoking the assumption of stationarity. Under stationarity, the multiple observations z(x1),..., z(xN) provide a kind of replication, i.e., any pair of observations z(xi), z(xi+h) are drawn from a bivariate process with the same distribution. This assumption allows us to infer information about D and ultimately model the random function Z(x).

The assumption of stationarity in its most general sense implies that the joint distribution of the random function at all locations does not depend on the absolute geographic location of the point samples, i.e. the models of spatially dependent variance (covariance structures) are the same over the entire sampled area. Formally, when a random function is strictly stationary, the joint distribution function at a set of N sample points, x1,..., xN, is invariant when the origin of x1,..., xN is translated. As we discuss below, the strict assumption of stationarity is not necessary in geostatistics, but a minimal assumption of stationarity in the mean (zero) and variance of the increments z(xi)-z(xi+h) is required.

The less serious breach of this assumption is a non-stationary mean, which can effectively be dealt with by using generalised covariances (Matheron, 1973) or the empirical best linear unbiased predictor with appropriate fixed effects (e.g., universal kriging; Meul and Van Meirvenne, 2003). A more serious violation of this assumption is when the variance is non-stationary and this is the focus of this paper. Non-stationary variance can take two forms, the variance may change as a function of space, i.e. the local variance of a soil property changes at different locations.Asecond form of non-stationary variance occurs when the distribution of the variance between scales (or spatial frequencies) changes in space. These changes in the scale-dependent distribution of variance imply a change in the autocorrelation function. Furthermore, variability of soil processes may well be non-stationary as a consequence of both changes in the variance and in the autocorrelation function.

CHARACTERISTIC SOIL PROPERTIES

It is easy to conceive of spatial processes in which it is not plausible to assume that the underlying covariance structures are stationary over the scales of interest. For example, pollution dispersion models in soils depend on certain characteristic soil properties such as texture, which can govern contaminant sorption or movement in the soil media. If empirical spatial covariances or empirical variograms are used to estimate texture across a landscape characterised by alluvial and glacial deposits as well as areas where erosion is prevalent, the assumption of an underlying random function which is stationary in the variance is implausible because of changes in local variation of texture. Consider measurements of soil carbon content on a transect that covers different parent materials, land use and vegetation. The characteristic spatial scales of variation in soil carbon content of clay soils under forestry may be largely determined by the structure of the forest canopy and management units for silvicultural production. If the transect also passes through a rapidly changing ecotone, from forest to wetland to grassland, then the pattern of variation in SOC may be of a larger range than in the woodland in this region of the transect, but with an important short range pattern in the wetland, reflecting carbon 'hot spots' due to microtopographic variation. If we compute a variogram for the whole transect,we assume, implausibly, that all this variation can be treated as a realisation of a single stationary random function. In fact, the variogram does not represent the variation in SOC anywhere in the forest because the variability changes so markedly. Other examples where stationaritymight be implausible are given by Lark (2006) and Sampson and Guttorp (1992).

The simplest approach to address changing variances is to partition the sampling region into segments within which stationarity is a more reasonable assumption (McBratney et al., 1991). There are also more complex techniques that can manage both forms of non-stationary variances, e.g. deformation of the original rectilinear coordinate space to obtain an alternative space in which stationarity is plausible (Sampson and Guttorp, 1992).

Currently the practitioner must infer where stationarity cannot plausibly be assumed by either observing directional variogram behaviour (possibly after removing a trend), by comparing variograms from portions of the sampling area or based on some underlying knowledge about the variable and processes being modelled. This could be facilitated by methods to test the plausibility of an underlying stationarity random function, and to identify how this assumption appears to fail.

Lark (2007) showed that it is possible to test the assumption of stationarity in variance using the discrete wavelet transform and Fuentes (2005) derived a spectral based test for stationarity.

However, the wavelet tests require that the data is obtained from a uniform sampling on a square grid or transect, which is not the most efficient approach to estimating the variogram.

We probably cannot assume that many soil properties are a realisation of a stationary random function because the variance and/or the auto correlation function appear to change. It is also the case that geostatistical methods that rely on the assumption of stationarity are increasingly being used in soil science and that methods exist that can alleviate some of the effects of non-stationarity.

However, there are currently few approaches to test for non-stationarity of spatial processes. In this paper, we propose an inferential method to test the plausibility of assumptions of stationarity in the variance given that the underlying model is a random function. The method tests for changes in the variance and auto correlations.

THE FORM OF STATIONARITY

Given the random function Z(x), strong stationarity occurs when at any finite number of points x1,..., xN, the joint distribution of Z(x1),..., Z(xN) is the same as that of Z(x1+h),..., Z(xN+h) where h is some displacement.A weaker form of stationarity is the assumption that a covariance function, Cov[Z (x+h), Z(x)] exists and depends only on h (equivalent to stationarity of the mean and variance only). This form of stationarity is known as second order stationarity. A still weaker form of the stationarity assumption was proposed by Matheron (1971, 1973). If the assumption of stationarity is made, not about the random function Z(x) but about the first order differences, Z(x+h)-Z(x), then this form of stationarity is known as the intrinsic hypothesis.

Under the intrinsic hypothesis, Z(x) is stationary if;

1. E[Z(x+h)-Z(x)]=0, for all x and h,
2. Var[Z (x+h)-Z(x)]=E{[Z(x+h)-Z(x)]2}=2?(h), for all x and depends only on h where ?(h) denotes the semivariance.

THE GLOBAL VARIOGRAM DATA

Themethod proposed in this Chapter computes local estimates of the variogram parameters, within a moving window, and compares their variability with that of the corresponding variograms for a realisation of a stationary random function with the same global variogram as the real data.

130 R. Corstanje et al./Geoderma 143 (2008) 123–132 few approaches currently available to determine whether the assumption of intrinsic stationarity of the variance is appropriate for a given set of data. The disadvantages of using moving windows are that they are computationally intensive for long range processes and the choice of the window size is somewhat arbitrary. The technique also requires fairly large datasets. The Bayesian spectral method developed by Fuentes (2005) generates similar inferences on non-stationary processes. It also requires some arbitrary decision because it is also a window-based approach and requires the selection of a weight function. This is not the case in the wavelet based approach presented by Lark (2006). However, the wavelet approach requires a dataset on a regular grid and may therefore be more appropriate for exhaustively sampled soil properties such as elevation or satellite imagery.

We have noted that our method is demanding of data.

However, large datasets on easy-to-measure variables (commonly from sensors) are now increasingly available in soil science (e.g. including remote sensing (de Bruin, 2000), electromagneticinduction (James et al., 2003), or yield monitoring (Dobermann and Ping, 2004). Our method may be of practical use in the analysis of such data. First, while densely sampled, these variables are often not regularly sampled and geostastistical methods may be needed to interpolate them, or to compute estimates at coarser scales by block kriging. These methods assume stationarity. Second, if this assumption of stationarity is implausible for an ancillary variable such as apparent electrical conductivity (ECa), then it is likely that it is implausible for a related soil variable (e.g. clay content). Information on how the stationarity assumption appears to fail for ECa might be used to plan sampling of the soil (e.g. to insure adequate sampling within different regions within which the assumption of stationarity is plausible). Finally, if the ancillary variable cannot be plausibly assumed to be stationary in the variance this will cause problems if it is to be used for cokriging direct measurements of a soil property. Again, in such circumstances field sampling of the soil might be planned so that the geostatistical analysis can be carried out separately in different subregions.

If our test suggests that stationarity is an implausible assumption, it will also indicate how this problem might be tackled in subsequent analysis while we look separately at the plausibility of stationarity in different aspects of the variability.

In the case where the spatial dependencies and ranges do not vary significantly but the variance does, then Hierarchical Generalised Linear Models (Lee and Nelder, 2001) might be suitable. In this approach, the variance is modelled as a function of some covariate(s). In the case where the variance seems to respond to localised perturbations, a robust estimator of the variogram (Lark, 2000) might mitigate the effects of local features of variation. Alternatively, if the variance changes as a result of distinct features in the landscape (such as the ecotypes cypress forest versus sawgrass marsh, or areas of high phosphorus soil content), then analysis could proceed by stratification for separate analysis. In the case where spatial dependencies and ranges appear to be non-stationary, but the local variance does not, non-linear transformations of coordinate systems (Sampson and Guttorp, 1992) might be appropriate. This method obtains local deformations of the coordinate space, a non-linear transformation of the coordinate space to a spacewhere the spatial structure is stationary.

Adaptations of this approach are increasingly being applied to spatial-temporal processes (see e.g., Schmidt and O'Hagan, 2003). Again, non-stationarity in the spatial dependence and/or range might be handled by stratification suggested by inspecting the local estimates of these parameters.

Index

❑❑❑